PATENTS HANDBOOK

PATENTS HANDBOOK

A Guide for Inventors and Researchers to Searching Patent Documents and Preparing and Making an Application

by FRED K. CARR

McFarland & Company, Inc., Publishers
Jefferson, North Carolina, and London

The present work is a reprint of the library bound edition of Patents Handbook: A Guide for Inventors and Researchers to Searching Patent Documents and Preparing and Making an Application, *first published in 1995 by McFarland.*

LIBRARY OF CONGRESS CATALOGUING-IN-PUBLICATION DATA

Carr, Fred K.
    Patents handbook : a guide for inventors and researchers to searching patent documents and preparing and making an application / by Fred K. Carr.
        p.   cm.
    Includes index.

    ISBN 978-0-7864-4321-5
    softcover : 50# alkaline paper ∞

    1. Patents — United States — Handbooks, manuals, etc.
    2. Patent searching — Handbooks, manuals, etc.   I. Title.
    T339.C347   2009
    608.773 — dc20   95-7366

British Library cataloguing data are available

Cover photograph ©2009 Shutterstock

Manufactured in the United States of America

McFarland & Company, Inc., Publishers
  Box 611, Jefferson, North Carolina 28640
   www.mcfarlandpub.com

To
Anna, Josh, Jake, Kory, and Delilah,
the reasons for all my efforts

CONTENTS

PREFACE

The purpose of the patent system is to promote the progress of science and useful arts. A patent grants to the inventor certain "exclusionary rights" for a period of years in return for the inventor disclosing his creativity. These exclusionary rights allow the inventor, or an organization, the right to exclude others from commercial exploitation of the invention while he or she has the right to make, use, or sell the invention. The exclusionary rights granted by the patent provide incentive for discovery, invention, development, and investment. Without this incentive, development and marketing of new technology would be greatly reduced.

In exchange for the above exclusionary rights, the patentee must by statutory mandate set forth an "enabling disclosure" of his invention in the patent specification. He must describe and teach the invention to the public such that the invention can be practiced without undue experimentation. The technology belongs to the public when the patent expires. The patent collection therefore contains a lot of prior art information, which for the most part is not fully utilized by the research community. A considerable amount of time and money is spent searching technical publications, but little time is spent searching patent documents. The failure to utilize this vast prior art collection often results in a waste of valuable research resources through the redoing of that which is already done, or worst yet, in the development of an intellectual property which belongs to another.

This book is designed for persons involved in research, prior art searching, and investing. These persons may have an interest in the patenting system for two reasons: many new products, processes, and uses can be patented which can lead to financial and other rewards for the inventor and his or her organization, and published patent documents are a rich source of prior art information.

The book, therefore, has the following objectives: to explain the general principles of the United States patent system, to explain the parts of the patent to the extent that they can be readily understood, to overview the process for filing and prosecuting patent applications, and to explain the different methods for searching patent documents for prior art information.

My interest in the patent system started many years ago as an academic exercise. Today I am chairman of a small electronics research and development company. Patents allow our smaller company to compete on an even keel with larger companies in the industry. The book is therefore written from the perspective of a business person. It is designed as a handbook to be used as a day-to-day reference in developing and protecting technology. The first few chapters overview the patenting process, the middle chapters relate to searching patent documents, and the final chapters discuss patent rights.

The major reference source is Title 35 of the United States Code, which contains statutes of the Patent Act of July 19, 1952, *Public Law 593, 82nd Cong., 2d sess., ch. 950,66 Stat. 792.* These statutes are cited as United States Code, Section 1, or the short citation 35 U.S.C. Sect. 1. This publication is obtainable from the Superintendent of Documents, U.S. Government Printing Office.

Another reference source is 37 Code of Federal Regulations. This code contains patent and trademark regulations as set forth by the Commissioner of Patents and Trademarks. Its short citation is 37 CFR 1.1.

Readers who want more detailed guidance than this book provides in the art of preparing and prosecuting patent applications should consult a treatise titled *Patent Preparation and Patent Practice*, edited by Irving Kayton, Patent Resource Institute, Inc., 2011 Eye Street NW, Washington, D.C. This is the premier treatise in the area of preparing and prosecuting patent applications.

Chapter 1

INTRODUCTION
TO PATENTS

WHAT IS A PATENT

A patent is an exclusionary right granted by a governmental entity. The concept behind the United States patent system is that the government grants statutory protection to an inventor in the form of exclusionary rights for a period of years in return for a disclosure of creativity by the grantee. The exclusionary rights granted by the patent are the right to exclude others from making, using, or selling the patented invention throughout the United States, and its territories and possessions for a period of 17 years. In exchange for these rights, the patent discloses and teaches technical knowledge relating to the invention. During the life of the patent, scientists and other inventors benefit from the disclosure of prior art information by avoiding repeating efforts to discover that which is already known. After the patent expires, the invention belongs to the public and anyone can make, use, or sell the invention without permission of the patentee.

The exclusionary rights granted by the patent to the patentee provide incentive for discovery, invention, development, and investment among inventors and corporations. Without this incentive, development and marketing of new technology would be greatly reduced. Neither corporations or individuals could afford to support research and development without these rights. Consider, for example, a prescription drug. Development of a prescription drug costs on average in excess of $150 million. No company would put up this amount of capital unless it thought it would get its investment back and make a profit. With patent protection of the drug, the patent rights enable the investor to prevent others from competing with him in the manufacture and sale of the drug during the life of the patent. This increases the probability that the investor will regain his investment and make a profit. At the same time the public reaps the benefit of the new drug.

The founders of the United States Constitution realized the importance

1

of promoting technology and thereby gave Congress the power to enact laws relating to patents. Article I, Section 8, of the Constitution specifies this authority and reads in pertinent part: "Congress shall have power . . . to promote the progress of science and useful arts by securing for limited times to authors and inventors the exclusive right to their respective writing and discoveries."

The first patent laws were enacted by Congress in 1790 and underwent general revision during 1952; this revision became effective January 1, 1953. These laws are reprinted in a United States government publication entitled *Patent Laws*, which is obtainable from the superintendent of documents. The United States Patent and Trademark Office, a division of the Department of Commerce, is assigned the responsibility for administering the laws relating to granting of patents.

A term that has often been erroneously associated with patents is the term *monopoly*. A patent is not a monopoly, but a contractual agreement between the inventor and the government. The government issues the patent which grants certain exclusionary rights to the inventor. In return for these rights, the inventor advances technology by disclosing his invention rather than concealing it from those working in the same area. As a result of this disclosure, science and humanity benefit. They may benefit from the invention itself, or they may benefit by the avoidance of repeated efforts to discover what has already been discovered. It is hard to imagine a patent so revolutionary that it would give control of the supply of a commodity or service in a given market, a condition necessary for a monopoly. Usually a patent encompasses only one product of many in a market. If the price of the patented product is excessive, consumers buy other products. If a patent were revolutionary in nature and encompassing in scope, however, it would have monopolistic properties.

A patentable invention is by statutory mandate a personal property. It is a property that can be developed, licensed, or sold (assigned). An assignment is the transfer of the right to make, the right to use, and the right to sell to the assignee. This is a total transfer of property. The patentee no longer has the exclusionary rights of the patent, but the assignee does. An assignment transfers undivided interest of the patent to the assignee for the full life of the patent.

A license is the transfer of something less than total transfer of property. A license gives the licensee the right to make, use, and sell the patented invention, but it does not give the licensee the right to exclude others from doing anything. The patent is still the property of the owner. In essence, the owner is giving the licensee the right to infringe the patent without any possibility of being sued for infringement. In contrast to the assignment, the license may be for a time period less than the life of the patent.

A patent for an invention is a grant from the government to an inventor and his heirs or assigns which grants certain exclusionary rights. These exclusionary rights include the right to exclude others from making, using, or selling the invention. The key phase in this statute is the "right to exclude." The patent does not give the patentee the right to make, use, or sell the invention per se, only the right to exclude others from doing so. If the patentee's rights infringe upon the rights of others, for example, a previously issued patent to another inventor, the patent does not give him the right to make, use, or sell the invention. In examining patent applications, the Patent Office does not make an absolute determination whether the invention sought to be patented infringes any prior patent. In the same light, if the patent violates any general laws that may be applicable, the patentee does not have the right to make, use, or sell the invention. In this regard, the key word of the patent right is the right to exclude others.

PATENTABLE INVENTIONS

Title 35, United States Code, Section 101, defines inventions patentable and is cited below:

> Whoever invents or discovers any new and useful process, machine, manufacture, or composition of matter, or any new and useful improvement thereof, may obtain a patent therefor, subject to the conditions and requirements of this title.

By this statute there are four classes of inventions: process, machine, manufacture, and composition of matter. To be patentable, these must be new and useful. In addition, any useful improvements in any of the four classes are patentable. The term "process" is defined in 35 U.S.C. Sec. 100 as a process, art, or method and includes a new use of a known process, machine, manufacture, composition of matter, or material. While not defined by statute, it has evolved over the years that the term "manufacture" refers to any article which is made. The term "composition of matter" refers to any chemical composition including new compounds and mixtures of ingredients.

Taken together, these four classes of inventions include practically everything made and the process for making it. However, by this statute and court interpretation of the statute, there is certain subject matter which is not patentable. Such subject matter includes naturally occurring articles, scientific principles, mental steps, printed matter, and methods of doing business. The Atomic Energy Act of 1954 also contains elements which exclude the patenting of inventions useful solely in the utilization of nuclear material or energy for atomic weapons.

In the United States, the inventor must apply for a patent. Title 35, United States Code, Section 111, states in pertinent part: "Application for patent shall be made by the inventor, except as otherwise provided in this title."

Title 35, United States Code, Section 116, states in pertinent part: "When an invention is made by two or more persons jointly, they shall apply for a patent jointly, and each sign the application." If a patent issues naming the wrong inventor(s), the patent would be invalid unless corrected.

While title to an invention must originate with the inventor, title is subject to vestment to another in some circumstances, the most common circumstance being that of an employed inventor. Much of the research and development done in the United States is done in an employer-employee relationship as with industrial concerns, academic institutions, clients of consultants, and the United States government. While patent applications filed for inventions resulting from this research must be filed in the name of the inventor(s), the employer does have vested rights to the invention. Such employer rights may result from common law rights of the employer or from a contractual agreement between employer-employee.

Common law, or nonstatutory law, holds that when an employer hires a person for the purpose of inventing or developing products, the employer is entitled to the fruits of creativity of the employee. Therefore even in the absence of an agreement between the employer-employee, the employer is entitled to assignment of an invention resulting from the work of the employee on the job. Most companies have a formal agreement with employees setting forth the obligation to assign inventions and other conditions that may be relevant.

Section 111 refers to certain exceptions whereby persons other than the inventor may file the patent application, and these exceptions are found in Sections 116, 117, and 118 of the Patent Code. These provide that if the inventor is dead, legally incapacitated, unwilling to execute the application, unavailable to execute the application, or cannot be found, the application may be made by an appropriate party on behalf of the inventor. In each of these situations, the true inventor must be designated on the application. When the inventor is dead, the application may be submitted by the administrator of the estate on the inventor's behalf. When an inventor refuses to execute the application or cannot be found, a person having a proprietary interest in the invention may submit the application on behalf of the inventor.

For an invention to be patentable, it must be novel, it must be nonobvious to those skilled in the pertinent art, and it must have utility. These are statutory requirements set forth in the Patent Code.

Section 102 sets forth the novelty requirements for patentability and is discussed in detail in Chapter 3. Most rejections of patent applications during the examination procedure are based on the first two paragraphs of Section 102, which provide that a valid patent cannot issue if:

> (a) The invention was known or used by others in this country, or patented or described in a printed publication in this or a foreign country before the invention thereof by the applicant for patent, or

(b) The invention was patented or described in a printed publication in this or a foreign country or in public use or on sale in this country more than one year prior to the application for a patent in the United States."

By this statute, if the invention has been described in a printed publication anywhere in the world or if it has been in public use or on sale in the United States before the *date of invention*, a valid patent cannot be obtained. If the invention has been described in a printed publication anywhere in the world, or if the invention has been in public use or on sale in the United States *more than one year* before the date on which the applicant filed his patent application in the United States, a valid patent cannot be obtained. Therefore if the inventor describes the invention in a printed publication, uses his invention publicly, or places the invention on sale, he must file his application within one year or lose patent rights.

Not all novel subject matter is patentable, however. Section 103 of the Patent Code provides that a patent may not be obtained even though:

the invention is not identically disclosed or described as set forth in section 102 of this title, if the differences between the subject matter sought to be patented and the prior art are such that the subject matter as a whole would have been obvious at the time the invention was made to a person having ordinary skill in the art to which said subject matter pertains.

Even though the subject matter sought to be patented has not been exactly described in the prior art, and even though there are some differences between the subject matter and the most nearly similar thing already known, a patent may be refused if these differences would be obvious to one skilled in the art.

The third statutory requirement for patentability is utility as set forth in Section 101 of the Patent Code:

Whoever invents or discloses any new and useful process, machine, manufacture, or composition of matter, or any new and useful improvement thereof, may obtain a patent therefore, subject to the conditions and requirements of this title.

The subject matter must have a useful purpose; the process must be able to accomplish its stated objective; the machine must be operational to perform its intended purpose; the article of manufacture must have a lawfully accepted purpose; and the chemical compound or composition must have practical utility in the chemical arts.

THE PATENT DOCUMENT

The patent grant itself is a single-page, red-ribboned document issued to the patentee and his heirs or assigns in the name of the United States under the seal of the Patent and Trademark Office. The ribbon document contains only the title of invention and the date of issue. Attached to and forming part of the patent are the specification and the drawings. The specification and drawings are the working parts of the patent.

The function of the drawings is to teach the invention to the public; they must show every feature of the invention specified in the claims. The specification is a written description of the invention. The drawings and the specification taken together must by statutory mandate provide an enabling disclosure of the invention that would "enable any person skilled in the art" to understand the invention.

At the end of the specification are the claims. The claims are one-sentence statements which point out and distinctly claim the subject matter which the patentee claims as his invention; they are in essence the property rights of the patent. The claims can be compared to the measurements of latitude, longitude, and distances in a real estate deed.

In the appendix to this book there is a sample United States patent entitled "Flow Signal Monitor for a Fuel Dispensing System," No. 5,361,216 issued to Walter E. Warn and Fred K. Carr. This patent was selected as an example not because of its particular merits, even though the author of this book prosecuted the application, but because it serves well to illustrate the component parts of a patent. This is a recently issued patent and follows the newer format used by the Patent Office for printing issued patents. Anyone not familiar with a patent might benefit from studying the format and components of the patent.

The first sheet of a patent document is the cover data sheet. This is a sheet prepared by the Patent Office for the purpose of summarizing bibliographic information about the patent. The patent number is at the top right corner of the page. This is a number given to each individual patent, and it serves to identify the patent. Below the patent number is the issue date. The life of the patent is seventeen years from this date.

To the left is the title of the patent. The function of the title is to direct persons with an interest in the subject area to the patent when searching prior art. Below the title is the name of the inventor(s) and his town of residence. The name of the assignee is printed just below the inventor.

The application number is the serial number assigned to the application which matured into a patent. Below the serial number is the date on which the application was filed.

United States patents are classified by both the International Patent Classification System and the United States Patent Classification System, which are listed. The bold print denotes the primary class and subclass; the regular print denotes the cross-reference classes and subclasses.

Also listed is the field of search showing the classes and subclasses in which the examiner conducted his prior art search for pertinent prior art along with references the examiner cited against the application that matured. Had there been printed publications cited, these would have been listed separately. The primary examiner and assistant examiner who examined the application and the patent practitioner who prosecuted the application are shown.

Finally, there is an abstract of the disclosure. Its purpose is to reveal to persons with an interest in the subject area the technical disclosure of the patent. Below the abstract is the number of claims in the patent and a representative drawing of the disclosure.

Following the cover data sheet are the drawings. The drawings show features of the invention and teach the invention to the public. With this particular patent, there are five sheets of drawings showing different schematic views, block diagrams, and flow charts of the invention. Most mechanical and electrical patents have drawings, while most chemical and process patents do not include drawings.

Following the drawings is the specification, which has two major functions: to provide an "enabling disclosure" of the invention and to define what is claimed by the invention; The specification, along with the drawings, must set forth the manner and process of making and using the invention, it must teach the invention to the public. To best organize the disclosure, the Patent Office prefers that the specification be divided into the following sections: field of the invention, background of the invention, summary of the invention, brief description of the drawings, and description of the preferred embodiment.

The field of the invention section is a statement which defines the field of art to which the invention pertains. It should be broad enough to cover all claims.

The background of the invention section describes the invention and any prior art related to the invention. It should discuss how the invention improves the prior art.

The summary of the invention section provides a summary and includes a description of the invention written in nontechnical terms. It should indicate the nature and substance of the invention along with the objectives of the invention.

The brief description of the drawings section describes the different views of the drawings, relating them to the written description of the invention in the specification.

The description of preferred embodiments section contains a written description of the invention. It must by statute describe the manner and process of making and using the invention and set forth the best mode contemplated by the inventor for carrying out his invention.

At the end of the specification are the claims. The claims set forth the metes and bounds of the invention. During examination of the patent application, the claims distinguish the invention from prior art; when the patent issues, the claims define the parameters of the invention. The claims are the operative part of the patent. It is the claims that set forth the property rights of the patent.

TYPES OF PATENTS

There are three types of patents issued by the United States Patent Office: technology (utility) patents, design patents, and plant patents. The most common type of patent is the utility patent; this is the type discussed in this book. The utility patent may encompass any new and useful process, machine, article of manufacture, or composition of matter, and any new and useful improvement thereof. The patent extends for a period of seventeen years from the date of.issue.

Design patents relate to a new, original, and ornamental design for articles of manufacture. The design patent protects only the appearance of an article; it does not protect the structure or substance of the article. The time extension for the design patent is determined by the fee paid to the Patent Office, and it can be obtained for three and one-half years, seven years, or fourteen years from the date of issue of the patent.

The plant patent relates to new and distinct varieties of plants, flowers, and trees. To be patentable, these must be asexually reproduced, that is, by means other than from seeds. This type of patent grants the patentee the right to exclude others from reproducing the claimed plants, flowers, and trees. The term of this patent is seventeen years from the date of issue.

PATENT INFRINGEMENT

If anyone without authority makes, uses, or sells a patented invention within the territory of the United States during the term of the patent, he infringes the patent. In cases of infringement, the patentee can sue for relief in federal court; he may ask that the infringement activity be stopped and may ask for damages. If the court rules in favor of the patentee, the court will usually award compensation for the infringement.

Questions of infringement are judged on the basis of the claims. Each claim of a patent is presumed valid when the patent issues, although the Patent Office does not make an absolute determination of validity. If the activity of the defendant falls within the scope of the claims, he is infringing the patent. During infringement proceedings, the defense usually asserts that his activity does not constitute infringement and that the patent allegedly infringed is invalid. As indicated, when a patent issues, it is presumed that the claims are valid. It is up to the party asserting invalidity to prove invalidity, which is decided by the court. The court may rule that the patent is valid and infringed, that the patent is valid but not infringed, that the patent is invalid but not infringed, or that the patent is invalid but would have been infringed if it were valid.

Infringement is a civil action, not a criminal action, and suits are initiated in federal court in the district where the infringing activity occurred. From the decision of district court, there is an appeal to the appropriate federal court of appeals and to the Supreme Court if certiorari is granted. The Patent Office has no jurisdiction in questions relating to infringement.

When a patented product is marketed, it should be so marked. The article should contain the word *Patent* and the number of the patent, or the word *Pat* and the number. The marking should appear on the item or on the container in which the item is enclosed.

Patent marking should be done by either the patentee selling the product or by any licensee selling the product. While it is not unlawful to market patented products without proper marking, it should be done for the following reason.

By statute, failure to properly mark a product as patented may result in a lost of infringement damages. When items are not marked, no damage can be collected by the patentee in an infringement suit unless the patentee can prove that he notified the infringer of his activity and the infringer continued this activity. In this case, damages can be collected from the date the infringer was notified.

When a product is made by a patented process, it should be marked to this effect. However, the marking should not imply that the product itself is patented, but only that the product was made by a patented process.

The phrases "Patent Pending" or "Patent Applied For" may be placed on items for which an application has been submitted. These phrases have no legal significance, however, and only show that an application has been submitted. The patent rights are afforded the article only when the patent issues. The one advantage that the use of this phrase might have is that it conveys to competitors that a patent has been applied for and they might be more hesitant to invest capital to set up production of the product.

False marking, or marking of an article as being patented when it is not, is against the law. Both civil and criminal liabilities can result from this activity. Therefore when the patent expires or when a patent is ruled invalid, the marking must be removed from literature within a reasonable time.

Patent marking differs from country to country. For example, in Canada the item carries the mark of "Patented" and the year it was patented, but usually not the number. Other countries use different markings; therefore, when items are patented and marketed in a particular country, one must use the appropriate marking for that country. Patent rights extend only in the countries in which the item is patented.

COPYRIGHTS, TRADEMARKS, TRADE SECRETS

Copyrights, trademarks, and trade secrets are often confused with the patent, although they are completely different entities and serve different purposes.

A copyright protects only the form of expression and not the subject matter of the writing. In essence, the copyright protects a writing from being copied. When a description of an article is copyrighted as a writing, this does prevent others from copying this description. This does not prevent others from writing their own description, however, or others from making, using, or selling this article. The only exclusionary right associated with the copyright is that of copying, whereas the patent has exclusionary rights of making, using, and selling.

A trademark is a symbol or word used by a manufacturer to identify his goods. The purpose of the trademark is to indicate the source of the merchandise and to distinguish this merchandise from that of other manufacturers. When the trademark is properly registered, it gives the right to prevent others from using the trademark or a similar mark that might confuse the public. It does not, however, give the right to prevent others from making, using, or selling similar merchandise under a nonconfusing symbol. Trademarks are potentially infinite in duration if properly registered. Many trademarks are centuries old.

A trade secret is any information which the owner considers valuable and does not disclose to the public. It is usually a process, a formula, or a manufacturing step used to produce an article of commercial value. The trade secret does in theory have some legal protection against unauthorized disclosure such as disclosure by former employees who were in a position of trust. If a person in a position of trust discloses a trade secret, the courts will sometimes award monetary compensation for the unauthorized disclosure if a competitor utilizes the disclosure. However, if the trade secret becomes common knowledge by independent discovery or by reversed engineering, the trade secret is lost. If it is not independently discovered, the trade secret has unlimited duration. Reversed engineering is not an illegal activity.

THE PATENT AND TRADEMARK OFFICE

The United States Patent and Trademark Office was established by Congress to perform the function of examining and issuing patents for the government. This office is a branch of the Department of Commerce and is located at Crystal Plaza, 2021 Jefferson Davis Highway, Arlington, Virginia. The head of the office is the commissioner of patents and trademarks, and the commissioner has several assistant commissioners. The work of the commissioner is subject to approval of the secretary of commerce.

The chief function of the Patent and Trademark Office is to examine patent applications to determine if the applicant is entitled to a patent under law and to issue such patent if he is entitled. The office also publishes issued patents and various other publications relating to patents and patent law, records assignments of patents, supplies patent copies, and maintains a search room for the public to use to examine issued patents and records. Over 100,000 applications are submitted to the Patent Office each year.

The Patent Office has about 1,350 patent examiners. The examiners are divided up into 15 "examining groups," of which 5 Groups are chemical examining groups, 5 are mechanical examining groups, and 5 are electrical examining groups. The examining groups are subdivided into "group art units," of which there are approximately 100. Each of these is assigned one or more general area of technology.

Each examining group is directed by a group director, and each group art unit within the examining group is headed by a supervisory primary examiner. Within the group art unit, there are a number of primary examiners who function independently and a number of assistant examiners who function under the supervisory primary examiner. Depending on the technology, examiners are generally well trained by education and experience in their subject area.

In addition to examining applications, the Patent Office publishes issued patents and other publications related to patents.

The Patent Office supplies copies of the five million plus patents which have issued over the years to the public for a fee. In addition, the Patent Office offers a service of mailing all future patents in a particular subclass as they issue when such subscription service is prearranged. By this method, one can follow a particular subject area.

The *Official Gazette* is the official journal of the Patent Office. It is published weekly, issuing on Tuesday, and contains a selected claim and drawing of each patent issuing that week. These are arranged according to the Patent Office classification system. In addition, the *Official Gazette* contains other matters such as a notice of patent and trademark suits, a list of patents for sale or license, an index of patents and patentees, and other information such as rules changes, etc. The *Official Gazette* is sold by subscription and single copy by the superintendent of documents.

The *Annual Index of Patents* is an index to the *Official Gazette* in two volumes, one an index of patentees and the other an index by subject matter of patents. It is available from the superintendent of documents. The Patent Office also publishes the *Annual Index of Trademarks*

Patents are classified by the Patent Office into a class and subclass according to their utility. The *Manual of Classification* which is also sold by the superintendent of documents, is a loose-leaf book containing a list of all classes and subclasses of this classification system. Classes and subclasses are constantly being reorganized, and substitute pages are issued from time to time. The *Classification Definitions* define in detail the subject matter assigned to each subclass. These are sold by subclass by the Patent Office.

OVERVIEW OF THE PATENTING PROCESS

For discussion, the subject of patents and patent law can be divided into three phases: the inventive phase, the preparation and prosecution of the patent application, and the exploitation of the patent rights. The inventive phase can be divided into the stages of conception, experimentation, and reduction to practice of the invention. Documenting these stages is an integral part of the patenting process because these stages determine inventorship, invention date, and diligence in reduction to practice. With reference to inventorship, the patent statutes require that the true inventor(s) be identified on the patent application; with reference to the date of invention, there are certain inchoate rights associated with the date of invention; with reference to diligence in reduction to practice, there are times as later discussed when the inventor must establish that he was diligent in reducing his invention to practice. Good record keeping is therefore a very important part of the inventive process. The inventor must have disclosure documents to substantiate the dates of the above events or he may lose patent rights.

Following reduction to practice, the next step is to consult a patent practitioner to determine if the subject matter is patentable. From an invention disclosure prepared by the inventor, the patent practitioner does a prior art search to determine if the subject matter is novel. If it appears that the subject matter meets the conditions for patentability, a patent application is usually submitted.

The patenting process is initiated by the applicant submitting a patent application to the Patent and Trademark Office. The application consists of a specification (including claims), drawings when appropriate, an oath or declaration by the applicant, and the filing fee. The components of the specification with claims and drawings were previously introduced in this chapter. The oath or declaration is in essence a statement by the applicant that he believes himself to be the original inventor of the subject matter in the application.

After a complete application has been filed, it is assigned to an examiner who specializes in the subject matter of the application. This examiner initially studies the application to determine if it follows the various Patent Office rules as set forth in 37 Code of Federal Regulations. One such rule is that each application contain only one invention. If the examiner deems that the application contains two inventions, he will request that the applicant select one invention for further study at this time and file for the other invention in another application.

If all formal matters are in order, the examiner proceeds to do a substantive examination of the application to determine if the subject matter meets the conditions for patentability. He does a prior art search of all issued United States patents in the pertinent art and searches through foreign patents, technical publications, etc., to determine if the subject matter is novel and nonobvious to those skilled in the pertinent art.

After the prior art search, the examiner makes a decision on the application and forwards this decision to the applicant, or actually to his patent practitioner, in writing. If the decision is adverse in any way, the applicant is given a time period to respond to the decision. The applicant may ask for reconsideration or reexamination of his application, with or without amendment. If he asks for reconsideration, he usually submits additional evidence relating to the application which is not evident in either the application or prior art. For example, he might submit evidence to stress a point concerning prosecution, or he might submit evidence to swear behind a prior art reference. To "swear behind" a prior art reference, an inventor must establish that he reduced the invention to practice in this country before the reference date or that he conceived the invention in this country and was diligent in reducing the invention to practice in this country from a date just before the reference date up to actual reduction to practice or to filing the application. On the other hand, he might amend his claims to overcome a prior art rejection by adding additional restrictions to his claims.

After considering the applicant's response and reexamination, the examiner's decision is again forwarded to the applicant in writing. Usually the second decision is a final decision; that is, this decision terminates the prosecution of the application on merits. There are certain circumstances in which the applicant is entitled to another reexamination, but this is quite rare in today's practice.

If the second decision is adverse to all or any part of the claims, the applicant may appeal the decision to the United States Patent Office Board of Appeals. This is a judicial body within the Patent Office to reconsider examiners' decisions. Adverse decisions from the Board of Appeals can be reviewed by either the Court of Custom and Patent Appeals or by the District Court of the District of Columbia, but not both. The Court of Custom and Patent Appeals is a constitutional court which considers the record of prosecution; no additional evidence may be presented during proceedings. Its decision can be appealed to the Supreme Court. The appeal to the District Court of the District of Columbia is a civil action suit, and it has appropriate appeals to the District of Columbia Circuit Court of Appeals and then to the Supreme Court.

The time involved during prosecution can vary greatly. Normally it takes about eighteen months for the prosecution stage, but it may vary from one year to five or more years. There are procedures to have new applications examined at an accelerated rate; the application is advanced out of its normal turn for examination by a petition from the applicant. This decreases the time involved in the prosecution stage considerably. On the other hand, if continuation applications are involved or if the applicant appeals or petitions an examiner's decision, several years may elapse from the initial filing date to the eventual issue date of the patent.

When claims in the application are found to be allowable, a notice of allowance is sent to the applicant along with a request for the issue fee, which is due within three months of the notice of allowance.

The third period associated with the patent is the exploitation period. This period begins when the patent issues and continues for the seventeen-year life of the patent. The patent itself creates no marketable product and puts no money in the bank. It does, however, give the patentee certain property rights which may with proper exploitation result in a money-making property. The essence of the patent right is the right to exclude others from commercial exploitation of the invention. The patentee may manufacture and sell the invention, may license others to do so, or may sell the entire property right to another.

Chapter 2

THE INVENTION

INTRODUCTION

An invention is generally defined as any new and useful process, machine, manufacture, or composition of matter, or any improvement therein. This chapter discusses the invention process and the various stages of the inventive process critical in transforming the invention into patent property. Also discussed is the importance of record keeping for establishing the dates on which certain events occurred during the inventive process. Dates of particular interest in patent protection are the date of conception of the invention, the date of reduction to practice of the invention, and the filing date of the patent application.

Flow chart 1 outlining the usual sequence of events from conception of the invention to the filing of patent application. The first stages of the process are conception, experimentation, and reduction to practice of the invention. These are pertinent in determining inventorship, date of invention, and eventual ownership of the patent rights if interference proceedings evolve.

In the United States, the inventor must be a real person and his name must appear on the patent application. This is in contrast to the practice in some foreign countries which allow corporations to apply for patents. In the United States, if the patent application is submitted in the name of the wrong inventor(s) and the patent issues, it would be invalid unless corrected. It is therefore important to determine the inventor accurately. Generally, the inventor is the person who first conceives of the invention provided he is diligent in reducing it to practice.

Determining the date of invention is an important aspect of patent law in the United States. There are certain potential rights associated with the date of invention. For an invention to be patentable, it must be new as defined by the patent statutes. One such statute relates to the invention date. In essence, this statute states that if the invention has been described in a printed publication anywhere in the world or if it has been in public use or on public sale in this country before the date the applicant made his invention, a valid patent

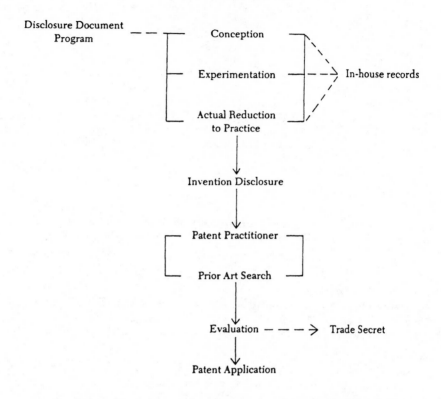

Flow chart 1. Process from conception to patent application.

cannot be issued. By this statute, it is possible to swear behind a reference as prior art if the date of invention comes before the reference date, even though the filing date of the application comes after the reference.

Another aspect of patent law associated with the invention date is the interference proceeding. Interference occurs in about one percent of the applications submitted to the Patent Office and occurs when two applications, an application and a patent, or two patents claim the invention. In essence, interference law states that the winning party is the party which first reduced to practice the invention unless the second party conceived the invention prior to the first party and the second party was diligent in reducing the invention to practice from a period just prior to the first party's conception. It can be seen that documentation of the date of conception, the date of actual reduction to practice, and proof of diligence in reduction to practice are very important aspects of the overall inventive process.

Following reduction to practice, the next step in the process is to obtain patent counsel to determine patentability. From an invention disclosure, a prior art search is conducted to determine if the invention meets the conditions for patentability. To be patentable, the invention must be novel, it must be nonobvious to those skilled in the art, and it must have utility. If the invention meets these requirements, the inventor usually submits a patent application. If it does not meet the requirements, the inventor may wish to keep the invention as a trade secret. The advantages and disadvantages of the trade secret are discussed in this chapter.

INVENTORSHIP

Title 35, U.S. Code, Section 111, explicitly states that an application for patent must be made by the inventor, and the completed patent application requires an oath by the applicant that he believes himself to be the first and original inventor. If a patent issues naming the wrong inventor, this is grounds for invalidation.

To define inventorship, it is helpful to establish a working definition of the term *invention*. Title 35, United States Code, Section 100, states that *invention* means invention or discovery. Title 35, United States Code, Section 101, defines *invention patentable* as a process, machine, manufacture, composition of matter, or any improvement therein. Therefore a working definition of an inventor is one who invents or discovers a new and useful process, machine, manufacture, or composition of matter, or any improvement therein. While this definition may seem simple in principle, in the modern corporate research setting it is often difficult to determine inventorship, especially in the case of joint inventorship.

Following are some of the general principles used in determining inventorship.

Conception, Experimentation, Reduction to Practice

The inventive process may be divided into the stages of conception, experimentation, and actual reduction to practice. Conception is a mental act of forming in the mind what an invention would be when reduced to practice. Conception usually precedes reduction to practice, although conception is never complete until reduction to practice is completed. Between conception and actual reduction to practice, there are experiments or physical steps for transforming the concept into a physical form. Experimentation may extend over a period of time; actual reduction to practice occurs at a given moment in time.

Conception is the process of forming in the mind the ideal of an invention to the extent that the invention would be operative when reduced to practice. An inventor must be involved in conception. Unless a person contributes to the conception of the invention, he is not an inventor. If a single person is responsible for conception, he is the sole inventor. If two or more persons jointly contribute to conception, there is joint inventorship.

The next step in the inventive process is experimentation toward actual reduction to practice. If the invention is a composition of matter or a process, the composition must be prepared or the process executed. If the invention is electrical or mechanical, it must be reduced to a physical form. In determining inventorship, being involved in experimentation does not make one an inventor. A person acting at the directive of another is not contributing to conception and is therefore not an inventor.

Actual reduction to practice occurs when the inventive concept is in a physical form which is workable. Establishing the workability of different types of inventions requires different degrees of testing. For a composition, various analytical and spectral tests must be completed. For a process, the process must be executed to the stage that the product is actually produced by the process. Electrical and mechanical inventions must be reduced to the stage that they perform the function for which they were designed. Actual reduction to practice occurs when the last test is performed to demonstrate workability. Therefore actual reduction to practice occurs at a given time, the date of which can be established with proper disclosure.

Most research departments are structured into the hierarchy of research director, research scientist, and research technicians. A research director may assign a project, but this does not make him a co-inventor; he must be involved in the actual conception of the invention to be a co-inventor. Research technicians who are acting at the direction of another person are not co-inventors, although they may be doing a lot of the work and contributing to the project. Conception is the key word in determining inventorship.

The question then arises whether the invention has to be actually reduced to practice before the patent application is filed. The answer is no. No patent statute requires that a working model of the invention exists before an application is filed; this would delay the filing of the application too long. Reduction to practice may occur in one of two ways: the concept may be tried and tested by normal experimentation (termed actual reduction to practice) or a patent application may be filed disclosing the concept sufficiently that one ordinarily skilled in the art can practice the invention (termed constructive reduction to practice). Indeed, most patent applications are filed before an actual working model is reduced to practice.

RESEARCH RECORDS

As discussed above, potential patent rights are associated with the date of conception, the date of reduction to practice, and the dates for establishing diligence in reduction to practice. To establish these dates legally, there must be research records or disclosure documents. One cannot simply say he conceived an invention, reduced it to practice, etc., on a given date without proper disclosure. To establish these events, the Patent Office requires prima facie evidence such as copies of records, notebooks, recordings, test data, etc. The Patent Office accepts such evidence prima facie. If court proceedings evolve, however, corroboration would be required.

Disclosure can take different forms. Companies involved in research most often use witnessed, in-house invention disclosures consisting of detailed records in hard-bound laboratory notebooks in which all stages of the inventive process are recorded. These records should consist of a written description of the event (conception, testing, etc.), and each page should be dated and signed by the inventor. In addition, these records should be read, dated, and signed by a collaborating witness.

Records should be kept up-to-date because it is very difficult to recall exactly what happened on a given date when time elapses from an event to the time it is recorded. Records should be kept in hard-bound notebooks; loose-leaf notebooks will not suffice. Data such as test results, analysis, recordings, etc., should be referenced in the notebook. These references should be witnessed and dated by the person performing the test. The quality of this notebook may determine if you are awarded patent rights or if your competitor is.

For individual inventors who do not want to disclose their inventive endeavors or for companies who want to enhance their documentation of conception date, the Patent Office has established the Disclosure Document Program. This is a program whereby the Patent Office will accept and keep on file for a limited time disclosure documents which serve as evidence of the date of conception of an invention.

The disclosure document itself is a written description of the invention, and it may include drawings or photographs relating to the invention. When submitting these documents to the Patent Office, the inventor must include a paper in duplicate stating that he believes himself to be the inventor and that he wants the enclosed documents placed on file in the Disclosure Document Program. In addition, the inventor must include a stamped, self-addressed envelope and a fee. The duplicate letter is returned to the inventor in the envelope provided, along with the filing number for future reference.

These documents are kept on file in confidence for a period of two years, after which time they are destroyed unless referred to in a patent application. In this case, they are retained for a longer period.

Some additional points should be made with regard to the Disclosure Document Program. These documents do not constitute a patent application, and any patent application relating to the disclosure will have the filing date of the application, not the filing date of the disclosure documents. These documents can be relied upon only as evidence of conception, not as evidence of reduction to practice or as evidence for establishing diligence in reduction to practice. Therefore, the Disclosure Document Program should not replace in-house record keeping.

DATE OF INVENTION

As previously indicated, there are certain statutory rights associated with the date of invention, such as swearing behind a prior art reference, interference proceedings, etc. There are times when the applicant needs to establish a date of invention prior to the filing date of his patent application. The earliest possible date of invention is the date of conception of the invention; the latest possible date of invention is the date on which the applicant files a complete patent application in the United States Patent Office or an international application. A date intermediate between these two which one might establish as a date of invention is the date of actual reduction to practice of the invention.

Establishment of a date of invention earlier than the actual United States filing date of the application depends on how well the stages of the inventive process have been documented and what occurred subsequent to the date of conception. When an inventor attempts to establish the date of conception as the date of invention, he must be able to establish that he was diligent in his efforts to reduce the invention to practice. He must have adequate disclosure to demonstrate a steady and earnest effort to reduce the invention to a physical form which is workable; that is, he must have records of experiments, records of supplies being ordered, etc., to demonstrate diligence.

If the inventor, or his workers, were not diligent from conception to actual reduction to practice, the earliest date of invention that he is entitled to is the date of actual reduction to practice. Reduction to practice refers to the actual construction of the invention in a physical form which is workable. For a machine, this is building the machine; for a composition of matter this is making and analyzing the compound or composition; for a process, this is production of a product by the process.

The latest possible date of invention is the date on which the applicant files a completed patent application in the Patent Office. Since the patent application must contain an enabling disclosure of the invention, when the patent issues, this serves as prima facie evidence that the invention was reduced to practice on the date the application was filed. The filing date of an application in this regard is referred to as constructive reduction to practice.

If conception or reduction to practice are used to establish a date of invention, these events must have taken place in the United States. When either conception or reduction to practice occurs abroad, the date of invention is by statute the filing date of the application in the United States or its effective filing date. This is set forth in Section 104 of the Patent Code cited below.

> In proceedings in the Patent Office and in the courts, an applicant for a patent, or a patentee, may not establish a date of invention by reference to knowledge or use thereof, or other activity with respect thereto, in a foreign country, except as provided in sections 119 and 365 of this title. Where an invention was made by a person, civil or military, while domiciled in the United States and serving in a foreign country in connection with operations by or on behalf of the United States, he shall be entitled to the same rights of priority with respect to such inventors as if the same had been made in the United States.

By this statute, foreign applications have an invention date based on the effective United States filing date. If the application does not rely on foreign priority, this is the date the application is filed in the United States. If the application relies on foreign priority, the invention date is based on the earlier filed foreign or international patent application.

As will be discussed more fully in later chapters, there are two circumstances under which a patent is given a filing date earlier than the actual filing date of the application from which it results. The earlier filing date is called the "effective filing date"; the date on which the application is filed in the United States is called the "actual filing date." An earlier effective filing date does occur when an application is given credit for "foreign priority," that is, credit for an earlier filed foreign or international patent application for the same invention. The other case of an earlier effective filing date may occur with the "continuing application," where credit is given for an earlier filed United States patent application or international application for the same invention.

SELECTING A PATENT PRACTITIONER

The next step in the process is to determine if the development meets the conditions for patentability. Usually at this point a patent practitioner is consulted. Determination of patentability, preparation of the patent application, and prosecution of the application in the Patent Office are responsibilities which require a knowledge of patent law as well as technical knowledge of the subject matter involved. While it is possible for the inventor to undertake this, it is not advisable unless he is knowledgeable in patent law.

Patent statutes give the Patent and Trademark Office the power to regulate persons conducting proceedings before it. The inventor may prepare and prosecute his own application without being registered with the Patent Office. Persons representing other inventors must be registered with the Patent Office, however, and such persons may be registered as "patent agent" or "patent attorney." Persons not recognized by this register of the Patent Office are not allowed by law to represent other inventors.

Considering the importance of scientific or engineering training in patent law, the Patent Office requires that all persons who wish to register take a special examination to demonstrate their level of competence. To qualify to take this examination, a person must have a college degree in science or engineering or the equivalent of such degree. An attorney-at-law who passes the Patent Office examination is registered as a patent attorney; a nonlawyer who passes the examination is registered as a patent agent.

In all respects the patent attorney and patent agent have equal privileges to practice before the Patent Office. They can both conduct proceedings within the Patent Office, including appeals to the United States Patent and Trademark Office Board of Appeals and interference proceedings before the Board of Patent Interferences.

In selecting a patent practitioner, it is important that the inventor select a person with expertise in the subject area of the patent application. The patent application is a technical document, and it takes technical knowledge for its preparation and prosecution. If a person with insufficient expertise in the subject area is selected, there is little likelihood that the patent obtained would adequately protect the invention.

The Patent Office maintains a directory of all registered patent attorneys and agents who have indicated an interest in representing inventors. While the Patent Office will not aid in the selection of an attorney or agent, it will respond to an inquiry whether a named person or firm is "reliable" or "capable."

When a patent practitioner is employed, the inventor executes a power of attorney which is filed with the Patent Office. Thereafter the Patent Office conducts proceedings with the patent practitioner, not the inventor. The inventor is free to contact the Patent Office concerning any aspect of his application, however. The attorney or agent can be removed by the inventor at any time by revoking the power of authorization.

INVENTION DISCLOSURE

The invention disclosure is a report prepared by the inventor to provide the patent practitioner with vital information on the invention. This is a most important report in that it forms the basis for conducting the patentability search and the patent application if the development meets the requirements for patentability. The preparation of the invention disclosure is a very important aspect of the inventive process.

The inventor is best qualified to describe what he considers to be his invention; he knows the origin of the invention, how the invention works, and what can be done with it. The inventor should take primary responsibility for properly completing the invention disclosure. Following is a summary of the information which should be included in the invention disclosure. The need for some of this information will become more apparent in Chapter 4, which discusses the patent application.

SUBJECT AREA OR TITLE OF THE INVENTION. This information assists the patent practitioner in organizing the prior art search and eventually in formulating a title for the patent application.

A LIST OF CONTRIBUTORS TO THE PROJECT. This list should indicate the full name and address of everyone involved in the project and the nature of their involvement in conception, experimentation, etc.

DISCLOSURE OF CONCEPTION. This should include the dates of the first discussion and any subsequent ones relating to the invention, including the type and location of records made during these discussions. In addition, the disclosure should include the date and location of the first written description of the invention and drawings in an organized form such as a laboratory notebook. If any documents have been submitted to the Disclosure Document Program, this should be indicated.

DISCLOSURE OF DEVELOPMENT. This should include the dates and names of persons doing development work on the invention, including the type and location of records relating to this work.

DISCLOSURE OF ACTUAL REDUCTION TO PRACTICE. This relates to the first successful test or demonstration of workability of the invention and should include dates and names of persons performing the test. The names, if any, of the persons who collaborated on the test should be included.

DESCRIPTION OF THE INVENTION. The patent application must contain an enabling disclosure of the invention, that is, a written description of the invention, how to make and use the invention, and the best method of practicing the invention. The description in the invention disclosure should be sufficiently detailed to teach the invention to the patent practitioner so that he can prepare the patent application.

PRIOR ART REFERENCES. These should include a list, with copies if available, of the most closely related prior art, including publications, patents, references in text books, etc.

DISCLOSURE TO OUTSIDERS. If the invention has been disclosed to anyone outside the company, this should be indicated. Such disclosure would include publications, abstracts, oral presentations, foreign applications, etc. The dates of any such disclosure should be indicated.

PUBLIC USE OR SALE. If the invention has been offered for sale, or displayed to the public in any way, this should be indicated including dates of disclosure.

PRIOR ART SEARCH

After studying the invention disclosure, a prior art search is conducted to determine if the subject matter meets the conditions for patentability. A prior art search should always be made before preparing the patent application. If prior art precludes the subject matter, the inventor is spared the expense of preparing the application. If the subject matter appears to be patentable, pertinent prior art from this search provides background information for preparing the application.

Finding and distinguishing the most pertinent prior art relevant to an invention is a very important aspect of the patenting process. Absolute novelty as defined by the patent statutes is required for patentability. If a patent issues without the examiner having considered certain pertinent prior art during the prosecution stage, this prior art can form the grounds for invalidation of the patent at a later time.

Prior art searching is often done in stages. The first stage of the search is referred to as the preliminary patentability search. The objective of this search is to determine the novelty of the subject matter. It consists of searching issued United States patents as well as relevant technical journals to determine if the subject matter or similar subject matter has been shown. Due to the time factor, this search is sometimes not as complete as that which will be made by the Patent Office. It is not uncommon for the Patent Office to reject claims in an application on the basis of prior art not found in the preliminary search.

Even after the application has been filed, however, there should be a continuous, on-going search for relevant prior art. The Patent Office has a procedure for the applicant to update his prior art statement as later discussed. This procedure is important because it is to the applicant's advantage to have all relevant prior art considered by the examiner; otherwise, at some point during the life of the patent it might be invalidated by prior art not considered during examination.

EVALUATION OF THE INVENTION

From the patentability report prepared by the patent practitioner, the decision is made whether the invention appears to meet the requirements for patentability. When it appears that these conditions are met, the inventor usually submits a patent application. If these conditions are not met, the subject matter may be kept as a trade secret. In some situations, the inventor may prefer to keep the subject matter as a trade secret even though it meets the requirements for patentability. A trade secret may or may not be patentable. There are some advantages of the trade secret over the patent, but in most instances the patent is of more value.

Patent Application Versus Trade Secret

PATENT APPLICATION. If a patent application is submitted and the patent issues, the patent offers several advantages over a trade secret. First of all, the patent grants certain exclusionary rights which continue for a period of years, and these rights are not lost if the subject matter is discovered by another during this period of time. The trade secret is, however, lost if another independently acquires the discovery.

The patent is more readily enforceable than a trade secret. The property rights of a patent are set forth in the claims. If these rights are infringed, the patentee has a document setting forth the property rights as defined by the claims and this document can be presented to the court. This is not the case with the trade secret because there is no public document defining the trade secret. The trade secret must be defined in court as later discussed.

In addition, the patent is more readily marketable. One can sell the patent rights, license these rights, or market the subject matter. A document setting forth property rights exists and can be presented to potential licensees or assignees. In the case of a trade secret, one cannot as readily go around explaining the subject matter to potential buyers, otherwise one would not have a trade secret long.

There are some disadvantages to the patent compared to the trade secret, however. The patent discloses to competitors one's area of interest and research, and this may stimulate competitor activity. It is common practice for companies to follow the research interest of competing companies by their patenting activity. The patent offers only finite protection in a subject area, and competitors can design their research activities around your patents. It is common practice for a company to analyze recent patents of other companies and "patent around" these patents.

Also, patents may be ruled invalid. While there is a presumption of validity when the patent issues, it may later be held invalid by the court system. When this happens, you are left with nothing.

TRADE SECRET. A trade secret can be generally defined as information which is valuable and which has been protected from public disclosure. In theory, the trade secret does have certain advantages over the patent. The trade secret can provide protection of the subject matter for an unlimited period of time, provided the secret is maintained, while patent protection lasts for only seventeen years. The trade secret enables one to avoid telling competitors information in the subject area in which one is working, whereas the patent discloses the subject matter which it protects, and this can stimulate competitor activity.

There are many disadvantages to the trade secret, however. The most prominent of these is the difficulty of maintaining a trade secret. During recent years there has been an increase in the movement of personnel, and this has compounded the difficulty of keeping trade secrets. Years ago it was not uncommon for an employee to spend his working life within the same company; now this is rarely the case. Employees move from company to company, and there is always the risk that some aspect of a trade secret will be disclosed to the next company.

If a trade secret is disclosed by a person in a position of trust and another company uses the information with the knowledge that it was disclosed by a person in a position of trust, the company can be sued. However, during litigation the owner of the trade secret has the burden of establishing that he does indeed have a trade secret. He must define the trade secret, convince the court that the information is of value, that the trade secret has been kept from the public, and that the secret was revealed by a person in a position of trust. This is difficult and usually not worth the legal cost. In contrast, the rights of the patent are set forth in the claims.

With the trade secret, there are no legal rights against the competitor who independently acquires the trade secret. If another discovers the subject matter encompassed by the trade secret, the subsequent inventor can use the subject matter at his discretion. In theory, this person could patent the subject matter and then sue for infringement if the original trade secret was still being used.

When a product made by a trade secret is marketed, it is often easy to "reverse engineer" the product to reveal information about the trade secret used in making the product. As analytical technology becomes more sophisticated, this approach makes many trade secrets less and less valuable. It should be emphasized that reverse engineering is not an improper method for discovering trade secrets. So all in all, there is a large degree of uncertainty associated with the trade secret.

Chapter 3

CONDITIONS FOR PATENTABILITY

INTRODUCTION

For an invention to be patentable, it must be novel, it must be non-obvious to those skilled in the pertinent art, and it must have utility. These are statutory requirements set forth in the United States Patent Code. Section 102 of the Patent Code specifies the novelty requirements for patentability along with certain conditions which may cause a loss of right to patent. Section 103 specifies the nonobvious requirement for patentability; that is, the invention must be sufficiently different from what was used or described in the past so that it would not be obvious to a person possessing ordinary skill in the pertinent art. The statutory requirement for utility is found in Section 101.

Any subject matter sought to be patented must be absolutely new; it must not have been known before either as a result of accident or design. The novelty requirement for patentability has been an integral part of patent law since the first statute of 1790. If the subject matter is known, it is in the public's possession and cannot be taken away from the public.

Not all novel subject matter is patentable, however. Section 103 of the Patent Code further requires that the difference between the development and preexisting technology be "such that the subject matter as a whole would not have been obvious at the time the invention was made to a person having ordinary skill in the area to which the subject matter pertains." This statute was added to the patent laws during the 1952 revision.

A majority of the rejections of patent applications are based on either lack of novelty (Section 102) or the obviousness of the invention (Section 103). The most frequently litigated aspect of patent law involves the novelty and nonobvious requirements for patentability. These two sections of the Patent Code are discussed in detail in this chapter.

NOVELTY

The novelty requirement for patentability along with certain conditions which may cause a loss of right to patent are set forth in Section 102 of the Patent Code cited below.

A person shall be entitled to a patent unless-

(a) the invention was known or used by others in this country, or patented or described in a printed publication in this or a foreign country, before the invention thereof by the applicant for patent, or

(b) the invention was patented or described in a printed publication in this or a foreign country or in public use or on sale in this country, more than one year prior to the date of the application for patent in the United States, or

(c) he has abandoned the invention, or

(d) the invention was first patented or caused to be patented, or was the subject of an inventor's certificate, by the applicant or his legal representatives or assigns in a foreign country prior to the date of the application for patent in this country on an application for patent or inventor's certificate filed more than twelve months before the filing of the application in the United States, or

(e) the invention was described in a patent granted on an application for patent by another filed in the United States before the invention thereof by the applicant for patent, or on an international application by another who has fulfilled the requirements of paragraphs (1),(2), and (4) of section 371(c) of this title before the invention thereof by the applicant for patent, or

(f) he did not himself invent the subject matter sought to be patented, or

(g) before the applicant's invention thereof the invention was made in this country by another who had not abandoned, suppressed, or concealed it. In determining priority of invention there shall be considered not only the respective dates of conception and reduction to practice of the invention, but also the reasonable diligence of one who was first to conceive and last to reduce to practice, from a time prior to conception by the other.

SUBSECTION (a). This subsection states that if before the applicant's discovery, the invention was known or used by others in this country or was patented or described in a printed publication in this or a foreign country, a valid patent cannot issue. An important element of this subsection is that it defines the novelty of the subject matter in terms of the date of the invention. This is unique among patent laws of the world. Only Canada, the Philippines, and the United States incorporate potential rights relating to the date of invention into their patent laws. Other countries base their laws around the filing date of the patent application. In these countries, certain events occurring between the invention date and the filing date of the application can bar an applicant from obtaining a patent. Such events would include a prior art reference. In the United States, this occurrence alone would not preclude the applicant's patent rights so long as he can establish a date of invention before the reference date of the prior art.

Another important aspect of this subsection is that it makes certain geographical distinctions with regard to knowledge and use by others and to being patented or described in a printed publication by others. If the invention was "known or used by others in this country," a patent is barred; if the invention was "patented or described in a printed publication in this or a foreign country," a patent is barred. Therefore, knowledge or use alone of the subject matter by others outside the United States, its territories, and its possessions does not preclude the issuing of a valid United States patent. However, if the subject matter is patented or described in a printed publication anywhere in the world, the applicant is precluded from obtaining a valid United States patent.

This geographic distinction is a matter of practicality. If subject matter is known or used in the United States, there is a reasonable chance that the applicant can become aware of it. If subject matter is known or used in a remote foreign country, chances are significantly less that the applicant would be aware of it. On the other hand, if the subject matter is published in a patent document or in a printed publication, typically it occurs in multiple-copy form and is accessible in libraries. It is therefore more reasonable to hold the applicant accountable for printed matter.

SUBSECTION (b). This subsection relates to certain events which occurred prior to the United States filing date of the patent application. It states that a valid patent cannot issue if the invention was patented or described in a printed publication in this or a foreign country or was in public use or on sale in this country more than one year prior to the United States filing date. The purpose of this statute is to prevent the inventor from being excessively slow in preparing and prosecuting his patent application, thereby being slow in disclosing his invention to the public.

The same geographic distinction exists with this subsection as with subsection (a). A patent or printed publication in this or a foreign country can preclude the issuing of a valid patent. The public use or public sale, on the other hand, must occur in this country. These events must have occurred more than one year prior to the date on which application was submitted.

It is noted that certain acts by the inventor himself may preclude his obtaining a valid patent by this subsection. If the applicant publishes the subject matter in this country or in a foreign country, he has a one year grace period to file the patent application. If this period elapses, the inventor is barred from a patent. If the inventor publicly uses or sells the subject matter in this country, he likewise must file the application within one year. The Supreme Court has held that "a single use for profit, not purposely hidden" constitutes public use.

When the last day of a subsection (b) bar falls on Saturday, Sunday, or a legal holiday in the District of Columbia, no bar arises if the application is filed on the next succeeding business day of the Patent Office. This is set forth in Patent Office Rule 7, which gives a one-year grace period plus one business day if the last day of the one year period falls on a weekend or holiday.

SUBSECTION (c). If the applicant abandons his invention, he is barred from obtaining a valid patent. Abandonment of an invention most often results from disclosing subject matter in an issued patent and not claiming the subject matter. Any matter disclosed, but not claimed, becomes public property when the patent issues. However, a decision by the United States Court of Customs and Patent Appeals gives a one-year grace period before abandonment is absolute. If the patentee files a reissue patent application within one year after his patent issues, presumption of abandonment is rebutted. The patentee may recover his patent rights to the disclosed but not claimed invention if it is claimed in the reissue patent.

Another form of abandonment which could preclude one from obtaining a valid patent under this subsection is intentional abandonment. Intentional abandonment would occur, for example, if a statement was put in a laboratory notebook clearly stating the research project was being stopped and that this particular invention was no longer being pursued. However, the act of slowing down an avenue of research for a while or putting it on the back burner does not constitute expressed abandonment. For intentional abandonment to occur, there must be a written statement so stating.

Abandonment of an application does not mean abandonment of the invention. If an application becomes abandoned for any reason, the applicant may refile another application claiming the same invention, provided no time bars have elapsed. Also, there are procedures for reviving abandoned patent applications in the Patent Office.

SUBSECTION (d). If more than twelve months prior to filling the United States application, the applicant filed a foreign application which issued into a patent prior to the filing of the United States application, a valid patent cannot issue. A bar arises here only when the United States applicant's foreign patent issues before his United States filing date and when the foreign application was filed more than twelve months before the United States filing date.

In filing foreign applications in most countries, this presents no problem since years elapse between filing the application and issue of the patent. However, in countries that issue inventors' certificates (Russia) or certificates of registration (Belgium) in lieu of patents, attention should be paid to this subsection since these may issue within twelve months.

SUBSECTION (e). If the invention was described in a patent based upon an application filed by another in the United States prior to the applicant's invention, a valid patent cannot be issued. The effective reference date of a United States patent as prior art is its filing date, not its issue date. The effective reference date of a printed publication as prior art is its publication date; the effective reference date of a foreign patent is its issue date. A United States patent as prior art is treated differently than other printed matter such as technical publications and foreign patents.

Since all patent applications filed in the Patent Office are maintained confidential and secret, this seems to generate a strange situation. The applicant is held responsible for prior art which he cannot become aware of by law. No information relating to a patent application can be given to anyone else without authority of the owner of the application.

If a patent issued on the same day that the application was filed, this situation would not exist, but, the prosecution of an application can continue for years. There are times, however, when an applicant becomes aware that another pending application is claiming the same patentable invention. Closely related art is examined within an art unit, and ideally, very closely related art is examined by the same examiner. If it is observed by an examiner that two or more applications are claiming the same invention, interference proceedings may be initiated by the examiner as later discussed. This proceeding determines the first inventor and therefore who is entitled to the patent.

The United States patent serves as prior art for all that it discloses, that is, the entire specification, including claims, descriptions, and drawings. It is prior art only if it issues; applications which never issue are not publicly disclosed.

SUBSECTION (f). This subsection states that if the applicant is not the inventor, a valid patent cannot be issued. Determination of inventorship was discussed in the previous chapter. In summary, the inventor(s) is the one who conceives the claimed invention in a complete and operative form. By this subsection, if the wrong inventor is named on the application and the patent issues, it will be invalid unless corrected.

SUBSECTION (g). The first sentence of this subsection states that a valid patent cannot issue if before the applicant's invention, the invention was made in this country by another who did not abandon, suppress, or conceal it. This sentence, in essence, provides that the first to reduce the invention to practice has priority unless he abandons, conceals, or suppresses the invention. For this sentence to apply, reduction to practice must have occurred in the United States. Abandonment in this subsection is the same as previously discussed in subsection (c). Concealment or suppression would occur, for example, if the invention were kept secret after reduction to practice. This sentence, in effect, is intended to penalize inventors who are excessively slow in disclosing their creativity.

The second sentence of this subsection qualifies the first sentence somewhat. This sentence provides that the date of conception, the date of reduction to practice, and diligence in reduction to practice should be considered when determining priority. This sentence provides that the party with priority is the first party to reduce the invention to practice unless another party first conceived the invention and was diligent in reducing it to practice from a time just before the first party's conception until the second party reduced the invention to practice. This subsection is the foundation of interference law, later discussed.

NONOBVIOUSNESS

Not all novel subject matter is patentable. The subject matter sought to be patented must be sufficiently different from that preceding it so that it would not be obvious to those skilled in the art. This requirement is set forth in Section 103 of the Patent Code cited below.

> A patent may not be obtained though the invention is not identically disclosed or described as set forth in Section 102 of this title, if the differences between the subject matter sought to be patented and the prior art are such that the subject matter as a whole would have been obvious at the time the invention was made to a person having ordinary skill in the art to which said subject matter pertains. Patentability shall not be negatived by the manner in which the invention was made.

Novelty is the first barrier that an application must overcome during the examination phase. Section 103 further imposes the requirement that the claimed invention be nonobvious at the time the invention was made to a person having ordinary skill in the art to which the subject matter pertains.

The problem of determining what is obvious and not obvious is a matter of judgment and is not well defined by the above stated statute. There are no set rules that one can follow in applying Section 103. The problem is further complicated by inconsistent court rulings relating to Section 103 over the years. As a result there is wide variation in the interpretation of this statute among the personnel in the Patent Office and an even wider variation in interpretation among the various circuits of the federal appeal courts.

Historically, the Patent Statutes of 1836 specified novelty as the only prior art requirement for patentability. In 1850, a court ruled for the first time (*Hotchkiss v. Greenwood*) that something more than novelty was required for patentability. For the next hundred years there were no common grounds for the numerous court decisions relating to this matter. Lack of invention was a term often used during this period. It was up to the judge in each case to determine if the subject matter was sufficiently different from that preceding it to be considered an invention and thus patentable.

Section 103 was added to the Patent Code during the 1952 revision for the purpose of clarifying this requirement. There is clearly a need for such a requirement; otherwise, the Patent Office would be flooded with patent applications claiming nonsubstantial subject matter. This objective has not been completely accomplished, however, because of the inconsistent interpretations given this statute as presently worded.

The question of obviousness under Section 103 is a problem that almost every inventor and patent practitioner must contend with. First, it may present a problem during the examination of the application: the examiner may reject claims on the grounds of obviousness. Also, it may present a problem during the exploitation phase. In almost every infringement suit, the defense asserts obviousness in an attempt to invalidate the patent that he is allegedly infringing. The reason for this is the wide range of interpretations given to Section 103 within the court system. The defense always has a chance of a decision of invalidation of the patent that he is allegedly infringing on the grounds of obviousness. Unlike Section 102, which has set rules stating the requirements for novelty, there are no set rules for applying Section 103.

This statute does not determine if the subject matter is an invention because it is novel and an invention if it meets the Section 102 requirements, but whether the invention is nonobvious to one having ordinary skill in the art. Therein lies the problem. The degree of ingenuity of an imaginary person having ordinary skill in the art is clearly a matter of judgment. Many things appear obvious in hindsight, but are in fact much more difficult beforehand. Many of our more useful inventions now seem simple and obvious, but, before their discovery they were not so simple or obvious or they would have been discovered earlier.

The major problem in applying Section 103 in every day patent law is the wide range of interpretations the court system has given Section 103. This is probably due in part to the fact that most judges do not have technical backgrounds and therefore do not fully understand discovery and invention. In particular, the Supreme Court has been inconsistent in its rulings on this statute. The Supreme Court is assigned the supreme authority on interpretation of a statute, and lower courts and Patent Office personnel attempt to follow its rulings. Following is an overview of the two most pertinent Supreme Court rulings on obviousness since the 1952 addition of Section 103.

The most quoted Supreme Court case relating to obviousness is the so-called "Trilogy." The Trilogy involved the obviousness issue of three individual patents: *Graham v. John Deere Company*, 148 U.S.P.Q. 459 (S. Ct. 1966); *Calman and Colgate-Palmolive Company v. Cook Chemical Company*, 148 U.S.P.Q. 459 (S. Ct. 1966), and the *United States v. Adams*, 148 U.S.P.Q. 479 (S. Ct. 1966). Without discussing each case individually, the Court wrote an overview of the rule that it applied in deciding these cases, and this is cited below. The disposition was written by Justice Clark.

> Under Section 103 the scope and content of the prior art are to be determined; differences between the prior art and the claims at issue are to be determined; differences between the prior art and the claims at issue are to be ascertained; and the level of ordinary skill in the pertinent art resolved. Against this background, the obviousness or nonobviousness of the subject matter is determined. Such secondary considerations as commercial success, long felt but unsolved needs, failures of others, etc., might be utilized to give light to the circumstances surrounding the origins of the subject matter sought to be patented. As indicia of obviousness or nonobviousness, these inquiries may have relevancy. (*Graham v. John Deere Co.*, 148 U.S.P.Q. 459 [S. Ct. 1966]).

In other words, the Court stated that all pertinent facts should be considered before a decision is made on the individual case as it relates to obviousness.

This was an opinion that provided a guideline which could be followed in everyday patent law.

It must be quickly added, however, that not all Supreme Court opinions have agreed with the above. Just three years after the Trilogy in *Anderson's — Black Rock, Inc. v. Pavement Salvage Co., Inc.*, 163 U.S.P.Q. (S. Ct. 1969), Justice Douglas wrote:

> A combination of elements may result in an effect greater than the sum of the several taken separately. No such synergistic result is argued here. It is, however, fervently argued that the combination filled a long felt want and has enjoyed commercial success. But these matters "without" invention will not make patentability. (*Anderson's Black Rock, Inc. v. Pavement Salvage Co., Inc.*, 163 U.S.P.Q. [S. Ct. 1969]).

This opinion in a sense counteracts the opinion in the Trilogy. It downplays the secondary considerations of long-felt want, commercial success, etc. Furthermore, this opinion emphasizes the abstract terms of combination and synergism, making it very difficult to interpret.

According to the above, a combination means all of the elements of the invention are known in prior art, whereas a noncombination means the invention has one or more new elements. How does one interpret this? Are not all composition of matter inventions combinations because there are only so many elements? In what way must new compounds or compositions be synergistic? With regard to mechanical devices, there are only so many belts, sprockets, chains, motors, etc. In what way must a combination of these be synergistic to meet the requirements of Section 103?

In determining a validity question relating to obviousness in federal court, the court may follow *Graham*, it may follow *Anderson's Black Rock, Inc.*, or it may establish its own guidelines. This often presents a problem for the patentee.

Where does the Patent Office stand during the examination of applications? The Patent Office does not require a demonstration of "synergism" per se during the examination of the application. The Patent Office does consider "secondary considerations" such as commercial success, long-felt but unsolved needs, failure of others, etc. Such evidence is accepted by way of Rule 132 affidavits later discussed. It is Patent Office policy to follow *Graham*, that is, consider all evidence presented in the case.

UTILITY

The statutory requirement for utility is found in Section 101 of the Patent Code printed in full below.

> Whoever invents or discovers any new and useful process, machine, manufacture, or composition of matter, or any new and useful improvement thereof,

may obtain a patent therefore, subject to the conditions and requirements of this title.

The language of this section clearly states that a process machine, manufacture, or composition of matter, or any improvement thereof must be useful in order to obtain a patent. If the claimed subject matter lacks utility, this is a basis for a rejection of the application.

The examiner assesses utility during examination of the application. Utility is usually apparent in subject matter relating to machines, articles of manufacture, and certain processes. The utility of a machine or an article of manufacture is usually self-evident from the disclosure in the patent application. The utility of a process is self-evident from the ability of the process to produce its stated objective.

Establishing utility of chemical compounds and compositions sometimes presents a problem. The Patent Office has established a standard whereby chemical compounds or compositions must have "practical utility in the chemical arts" to meet the utility requirement. Scientific research is insufficient basis for utility; being an intermediate in a synthetic pathway for another compound is insufficient basis for utility. Often an applicant is required by the Patent Office to submit test data in addition to the application to aid the examiner in assessing utility of chemical compounds or compositions.

The one class of chemical compounds and compositions that has caused examiners the greatest difficulty in assessment of utility is chemicals claiming pharmaceutical utility. Since it is very difficult to predict biological activity from chemical structure, the Patent Office is reluctant to allow claims encompassing compounds alleging pharmaceutical utility. Often the examiner will require extensive test data relating to utility by affidavit to aid in his assessment.

Considering the unpredictability of structure biological activity relationships, the Patent Office has established in the *Manual of Patent Examiner Procedure*, Section 608.01(P) the following principles relating to pharmaceutical utility that are to be followed in considering matters relating to the adequacy of disclosure of utility in drug cases:

> (1) The same basic principles of patent law which apply in the field of chemical arts shall be applicable to drugs, and
> (2) The Patent Office shall confine its examination of disclosure of utility to the application of patent law principles, recognizing that other agencies of Government have been assigned to the responsibility of assuring conformance to the standards established by statute for the advertisement, use, sale or distribution of drugs.

The reasoning behind these guidelines is that the Patent Office does not want to imply a useful pharmaceutical utility for any allowed compounds or compositions. Therefore, the patentee cannot imply pharmaceutical utility on the basis that the compound was patented with this particular utility.

Chapter 4

THE PATENT APPLICATION

INTRODUCTION

In order to obtain a patent for an invention, the inventor must submit to the commissioner of patents and trademarks an application containing the following parts: (1) a written document comprising a specification (description and claims), (2) an oath or declaration, (3) drawings in cases where drawings are appropriate, and (4) the required filing fee. An application which contains these elements is regarded by the Patent Office as a "complete" application, and it is given a filing date and application serial number. A filing date and application number are assigned only on the date on which all of these elements are in the Patent Office.

Applications which are complete with regard to parts, but do not meet Patent Office standards because they are nonpermanently printed, nonreproducible, printed on both sides of the paper, etc., are regarded as "informal" by the Patent Office and are given neither a filing date or application number. In such cases the applicant is informed that the application is too informal to be examined, and the Patent Office allows six months for these informalities to be corrected. If the defects are not corrected, the application is returned along with any filing fees that the applicant has submitted.

Applications are numbered by arrival sequence, that is, the date on which a complete and acceptable application is received in the Patent Office. The applicant is thereafter informed of his application serial number and filing date. When an incomplete or informal application is denied its original filing date, Patent Office rule 183 provides that a petition to waive the rules may be filed. Such petitions are usually allowed by the commissioner only when it can be shown that a proper application could not have been submitted at that time and that irreparable damage would result from a denial of the filing date.

Following is a discussion of the sections of a patent application.

SPECIFICATION

Title 35, United States Code, Section 112, relates to the specification and is printed below.

The specification shall contain a written description of the invention, and of the manner and process of making and using it, in such full, clear, concise, and exact terms as to enable any person skilled in the art to which it pertains, or with which it is most nearly connected, to make and use the same, and shall set forth the best mode contemplated by the inventor of carrying out his invention.

The specification shall conclude with one or more claims particularly pointing out and distinctly claiming the subject matter which the applicant regards as his invention. A claim may be written in independent or, if the nature of the case admits, in dependent or multiple dependent form.

Subject to the following paragraph, a claim in dependent form shall contain a reference to a claim previously set forth and then specify a further limitation of the subject matter claimed. A claim in dependent form shall he construed to incorporate by reference all the limitations of the claim to which it refers.

A claim in multiple dependent form shall contain a reference, in the alternative only, to more than one claim previously set forth and then specify a further limitation of the subject matter claimed. A multiple dependent claim shall not serve as a basis for any other multiple dependent claim. A multiple dependent claim shall be construed to incorporate by reference all the limitations of the particular claim in relation to which it is being considered.

An element in a claim for a combination may be expressed as a means or step for performing a specified function without the recital of structure, material, or acts in support thereof, and such claim shall be construed to cover the corresponding structure, material, or acts described in the specification and equivalents thereof.

The specification has two major functions: to provide an "enabling disclosure" of the invention and to define what is claimed by the invention. The enabling disclosure provides a description from which the public can learn what the invention is, improve on the invention during the life of the patent, and practice the invention when the patent expires. The specification must be written in such clear and exact language as to enable any person skilled in the art to which the invention relates to make and use the invention. In addition, the specification must set forth the best method contemplated by the inventor for carrying out the invention at the time he made the invention. The best method provision prevents inventors from applying for a patent while at the same time concealing from the public information about the invention.

The claims are statements at the end of the specification which point out and distinctly claim the subject matter of the invention. The claims distinguish the invention over prior art and recite all essential features of the invention. The claims are the operative part of the patent. Both novelty and patentability of the invention are judged by the claims during the prosecution phase of the application. If the patent issues, questions on infringement are judged by the courts on the basis of the claims; the claims are the property rights of the patent.

Although the Patent Office allows a great deal of discretion by the patent practitioner in formulating the specification, the Patent Office has set forth rules and suggestions as to what an "ideal" application might look like. The preferred format by the Patent Office contains the following sections.

1. Title of the Invention
2. Cross-References to any Related Application
3. Field of the Invention
4. Background of the Invention
5. Summary of the Invention
6. Brief Description of the Drawings
7. Description of the Preferred Embodiments
8. Claims
9. Abstract of the Disclosure

Title of the Invention

The title of the invention should be as short and specific as possible. It should be descriptive enough that when the patent issues, the title will direct other persons with an interest in the subject matter to the patent when searching issued patents or abstracts of issued patents. If an unsatisfactory title for this purpose is submitted by the applicant, the examiner can change the title by amendment when the application is allowed.

Field of the Invention

The background section of the application should contain a statement on the field of the invention and a description of prior art related to the invention. The statement on the field of invention defines the field of art to which the invention pertains. This statement should be broad enough to cover all claims in the application, but it should be selectively worded because it influences the art group to which the application is assigned for examination.

Background of the Invention

This section is a description of prior art related to the invention. This is a very important part of the application. First, there is a duty of disclosure of prior art on the part of the inventor and others involved with the application, and second, the validity of the patent resulting from the application depends on the record of prior art considered during examination. It is to the inventor's advantage to have prior art considered and made of record during the examination phase.

If the resulting patent is commercially successful, there is a probability that the patent will be infringed. The defense for the alleged infringer will probably assert that the patent is invalid because it does not meet the novelty and nonobviousness requirements for patentability, and the alleged infringer will present prior art from the four corners of the earth. The point is that if the presented prior art is the same as that considered by the examiner, the court will maintain validity in most cases. If the presented prior art is different and pertinent, chances of invalidity are significantly increased. Since nothing is more nonproductive than an invalid patent, all known prior art should be presented for consideration by the examiner.

There is a duty of disclosure of prior art on the part of the inventor, the attorney or agent who prepares the application, and anyone involved in the prosecution of the application or research of the invention. Failure to make such disclosure may result in a fraudulent patent. The duty of disclosure is set forth in 37 CFR 1.56 cited below in pertinent part:

> (a) A duty of candor and good faith toward the Patent and Trademark Office rests on the inventor, on each attorney or agent who prepares or prosecutes the application and on every other individual who is substantively involved in the preparation or prosecution of the application and who is associated with the inventor, with the assignee or with anyone to whom there is an obligation to assign the application. All such individuals have a duty to disclose to the Office information they are aware of which is material to the examination of the application. Such information is material where there is a substantial likelihood that a reasonable examiner would consider it important in deciding whether to allow the application to issue as a patent. The duty is commensurate with the degree of involvement in the preparation or prosecution of the application.
>
> (b) Disclosures pursuant to this section may be made to the Office through an attorney or agent having responsibility for the preparation or prosecution of the application or through an inventor who is acting in his own behalf. Disclosure to such an attorney, agent or inventor shall satisfy the duty, with respect to the information which is not material to the examination of the application.

This rule generally states that any information that is material to the examination of the application should be disclosed to the Patent Office when there is a likelihood that the examiner would consider the information while examining the application. The two major sources of prior art are the literature search the inventor did for research purposes and the patentability search the patent practitioner conducted to determine patentability. From this list, the decision is made about which references to disclose and which not to disclose. The best person to decide this is the patent practitioner. By human nature, inventors are often unwilling to believe that anyone else could have possibly come up with an ideal even similar to their invention, and inventors therefore tend to withhold prior art. For this reason, the patent practitioner is usually in the best position to make this decision.

Once the decision has been made that a reference is material and should be cited, the decision must be made about the best method of informing the examiner about the prior art. There are two methods; the reference can be included in the specification or the reference can be made part of the prior art statement. The prior art statement is provided for in 37 CFR Sects. 1.97, 1.98, and 1.99. Section 1.97 sets forth the format and the time for filing the prior art statement and is cited below:

> (a) As a means of complying with the duty of disclosure set forth in § 1.56, applicants are encouraged to file a prior art statement at the time of filing the application or within three months thereafter. The statement may either be separate from the specification or may be incorporated therein.
> (b) The statement shall serve as a representation that the prior art listed therein includes, in the opinion of the person filing it, the closest prior art of which that person is aware; the statement shall not be constructed as a representation that a search has been made or that no better art exists.

This rule provides that the prior art statement may be either separate from the specification or incorporated within it. If separate, the statement should be filed at the time of filing the application or within three months thereafter. The rule also states that the prior art listed in the application is to be the closest prior art of which the person is aware; however, it goes on to say "the statement shall not be construed as a representation that a search has been made or that no better art exist."

Rule 98 sets forth the content of the prior art statement and is cited below:

> (a) Any statement filed under 1.97 or 1.99 shall include: (1) A listing of patents, publications or other information and (2) a concise explanation of the relevance of each listed item. The statement shall be accompanied by a copy of each listed patent or publication or other item of information in written form or of at least the portions thereof considered by the person filing the statement to be pertinent.
> (b) When two or more patents or publications considered material are substantially identical, a copy of a representative one may be included in the statement and others merely listed. A translation of the pertinent portions of foreign language patents or publications considered material should be transmitted if an existing translation is readily available to the applicant.

Rule 99 provides for updating the prior art statement and is cited below:

> If prior to issuance of a patent an applicant, pursuant to his duty of disclosure under sect. 1.56, wishes to bring to the attention of the Office additional patents, publications or other information not previously submitted, the additional information should be submitted to the Office with reasonable promptness. It may be included in a supplemental prior art statement or may be incorporated with other communications to be considered by the examiner. Any transmittal of additional information shall be accompanied by explanations of relevance and by copies in accordance with the requirements of Sect. 1.98.

The Patent Office requests that applicants use the form PTO-1449, "List of Prior Art Cited by Applicant" when submitting a prior art statement under 37 CFR 1.97, 1.99. Each reference, when considered by the examiner, will be printed on the issued patent as a prior art reference. This form can be obtained from the Patent Office.

Summary of the Invention

Patent Office Rule 73 relates to the summary of the invention and states:

> A brief summary of the invention indicating its nature and substance, which may include a statement of the object of the invention, should precede the detailed description. Such summary should, when set forth, be commensurate with the invention as claimed and any object recited should be that of the invention as claimed.

This section should include a description of the invention in nontechnical terms. This description could become very important if the patent is involved in litigation since judges are not well trained in technical areas and do not fully understand the claimed invention. If the patent contains a section describing the invention in nontechnical language, there is a better chance that the judge will comprehend the invention.

Brief Description of the Drawings

If the patent application contains drawings, the specification should contain a section describing the drawings. If the drawings include different views of the invention, this section should refer to different views by appropriate markings such as reference letters or numerals. This section should amplify what the drawings represent and how these fit into the scheme of the invention.

Description of the Preferred Embodiments

By statutory requirement, the specification must contain an enabling disclosure of the subject matter which would enable anyone skilled in the pertinent art to practice the claimed invention without experimentation, and it must describe the best mode contemplated by the inventor at the time of filing the application for carrying out the invention. It is within this section of the application that such detailed description is set forth, although the entire application is considered by the Patent Office when determining if the application contains an enabling disclosure.

The enabling disclosure requirement is set forth in the first paragraph of Section 112 of the Patent Code cited below in pertinent part:

The specification shall contain a written description of the invention, and of the manner and process of making and using it, in such full, clear, concise, and exact terms as to enable any person skilled in the art to which it pertains, or with which it is most nearly connected, to make and use the same,and shall set forth the best mode contemplated by the inventor of carrying out his invention.

It is emphasized that the disclosure requirement can rely only on the disclosure contained in the original application on the date it is filed in the Patent Office. New matter for the purpose of amending the specification to meet the disclosure requirement is not allowed by statute. If matter necessary for disclosure is omitted at filing, it can be added only by filing a new application or by filing a continuation-in-part application. Both of these with regard to the new matter rely on their filing date that is, the original filing date is lost. If such disclosure material is omitted from the original application, it is the fault of the patent practitioner. Unfortunately, it is the inventor who suffers the consequence of losing the filing date. If a statutory bar arises during this time, patent rights will be lost.

The style and detail of this section of the application depends on the class of invention being described. In mechanical and electrical applications, this description usually involves identifying the elements of the invention with reference to the drawings, describing how these elements fit together to form the invention, and describing how these elements function together to make the invention work. With mechanical and electrical applications, the drawings are very instrumental in the enabling disclosure. The drawings are relied on to explain the details of the invention and to teach the invention to the public.

Chemical cases usually do not include drawings. Therefore in chemical applications, this section must include a detailed description of the method for making the compound or composition. If the application is a chemical process, it must describe fully each step in the process. The description for a compound should include starting materials, reaction conditions, temperature, catalyst, pressures, isolation procedures, etc. For compositions, this description should include ingredients, reaction conditions, etc. For chemical processes, which may or may not include drawings, the description should describe all steps in the process as well as suitable materials used in the process. Since there are ordinarily no drawings in chemical cases to rely upon for detail, this written description must be detailed enough to describe the invention, the manner and process of making and using the invention, and the best mode contemplated by the inventor for carrying out the invention.

It is common practice in chemical cases to include working examples in the application. Since an application for new compounds usually encompasses many compounds, it would be unrealistic to detail the synthetic scheme of each compound claimed. Therefore, working examples of representative compounds are described. The number of examples used in the application depends on the nature of the compounds being claimed. Working examples of work already done are written in past tense. Working examples for compounds not yet synthesized are written in present tense or future tense.

When the case relates to an improvement in a process, machine, manufacture, or composition of matter, the specification must point out the part or parts which constitute the improvement. The description should explain how the invention is an improvement over prior art and how the elements are involved in the improvement.

A unique situation occurs in cases relating to microorganisms. No matter how extensive the description, it cannot put the organism in the public's possession, a requirement that must be met to satisfy Section 112. To satisfy this requirement, the Patent Office requires that newly discovered microorganisms for patent be placed in a depository. The deposit must be made no later than the effective United States filing date, and the culture must be in a depository capable of continuing a permanent culture. During the examination phase of the application, only Patent Office personnel have access to the deposit. If the patent issues, the public then has access to the culture.

The Patent Office allows certain material to be incorporated into the specification by reference. This is an attempt to keep the specification from being too bulky. If the material incorporated by reference is necessary either for providing an adequate disclosure or for providing support for the claims, the material is labeled "essential" by the Patent Office. The Patent Office allows the incorporation of essential material by reference only to issued, or allowed, United States patents. Other nonessential material may also be incorporated by reference, including issued United States and foreign patents, earlier filed, commonly owned, United States applications, or other nonpatent documents as printed publications.

Claims

At the end of the specification are the claims, which particularly point out and distinctly claim the subject matter as described in the previous section. Usually there is more than one claim. There is no limit on the number of claims a patent can have, provided they differ substantially from each other and are not unduly multiplied. The claims may be in independent form or dependent form, where dependent claims refer back to and further limit another claim or claims. Any dependent claim which refers to more than one other claim is called a multiple dependent claim. A multiple dependent claim can not serve as a basis for any other multiple dependent claim. Dependent claims include all the limitations of the claim incorporated by reference. A multiple dependent claim incorporates by reference all the limitations of each of the particular claims from which it depends.

The claims must conform to the invention as described in the previous part of the specification. The terms and phrases used in the claims must find clear support or antecedent basis in the description so that the meaning of the terms in the claims can be understood by reference to the written description.

In the case of an improvement claim, the independent claims are written in the following order: a preamble comprising a general description of all the elements or steps of the claimed combination which are conventional or known, a phrase such as "wherein the improvement comprises," and those elements, steps, and their relationship which constitute the portion of the claimed combination which the applicant considers as the new or improved portion.

The claims are further discussed in the following chapter.

Abstract

Following the claims, there is an abstract of the disclosure. The purpose of the abstract is to relay to interested persons the gist of the technical disclosure in the patent from a cursory inspection of the abstract. It is usually written in nontechnical terms and serves not to interpret the scope of the claims, but to direct people with interest to the patent.

OATH

The Patent Statute requires that an oath or declaration be submitted with the patent application. The requirement for the oath is set forth in Section 115 of the Patent Code, cited below:

> The applicant shall make oath that he believes himself to be the original and first inventor of the process, machine, manufacture, or composition of matter, or improvement thereof, for which he solicits a patent; and shall state of what country he is a citizen. Such oath may be made before any person within the United States authorized by law to administer oaths, or,when made in a foreign country, before any diplomatic or consular officer of the United States authorized to administer oaths, or before any officer having an official seal and authorized to administer oaths in the foreign country in which the applicant may be, whose authority shall be proved by certificate of a diplomatic or consular officer of the United States, and such oath shall be valid if it complies with the laws of the state or country where made. When the application is made as provided in this title by a person other than the inventor, the oath may be so varied in form that it can be made by him.

In addition to these statements, the Patent Office further requires other allegations be made in the oath as will be developed later in this section.

Present Patent Office rules allow a declaration to be used in lieu of an oath. An oath is a sworn document and must be signed before an appropriate officer such as a notary; a declaration need not be notarized, but it must contain a warning to the declarant that willful false statements and the like are punishable by fine or imprisonment or both. The commissioner allows the filing of a signed declaration instead of a notarized oath in most papers submitted to the Patent Office. Since the declaration is more convenient, it is more commonly used.

Below is a combined declaration and power of attorney used to accompany an original application as cited from 37 CFR 3.16a. It is not necessary to include the power of attorney with the declaration, but it usually is included at this point as a matter of convenience. It saves another signature by the inventor.

As a below named inventor, I hereby declare that:

My residence, post office address and citizenship are as stated below next to my name; that

I verily believe I am the original, first and sole inventor of the invention entitled:

_____

described and claimed in the attached specification; that

I do not know and do not believe the same was ever known or used in the United States of America before my or our invention thereof, or patented or described in any printed publication in any country before my or our invention thereof or more than one year prior to this application, that the same was not in public use or on sale in the United States of America more than one year prior to this application, that the invention has not been patented or made the subject of an inventor's certificate issued before the date of this application in any country foreign to the United States of America on an application filed by me or my legal representatives or assigns more than twelve months prior to this application, that I acknowledge my duty to disclose information of which I am aware which is material to the examination of this application, and that no application for patent or inventor's certificate on this invention has been filed in any country foreign to the United States of America prior to this application by me or my legal representatives or assigns. Except as follows:

I hereby appoint the following attorney(s) and/or agent(s) to prosecute this application and to transact all business in the Patent and Trademark Office connected therewith:. Address all telephone calls to: _____ at telephone no.: _____

Address all correspondence to: _____

I hereby declare that all statements made herein of my own knowledge are true and that all statements made on information and belief are believed to be true; and further that these statements were made with the knowledge that willful false statements and the like so made are punishable by fine or imprisonment, or both, under Section 1001 of Title 18 of the United States Code and that such willful false statements may jeopardize the validity of the application or any patent issued thereon.

Full name of inventor: _____

Signature: _____

Residence: _____

Citizenship: _____

Post Office Address: _____

The declaration must state whether the applicant is a sole or joint inventor. The usual practice is for joint inventors to execute the same declaration, although it is acceptable for joint inventors to execute separate declarations so long as they make reference to each other.

DRAWINGS

An application for a patent is required to contain drawing(s) of the invention whenever the subject matter merits drawings. The purpose of the drawings is to teach the invention to the public when the patent issues. The statutory requirement for drawings is set forth in 35 U.S.C. Sect. 113 cited below.

> The applicant shall furnish a drawing where necessary for the understanding of the subject matter sought to be patented. When the nature of such subject matter admits of illustration by a drawing and the applicant has not furnished such a drawing, the Commissioner may require its submission within a time period of not less than two months from the sending of a notice thereof. Drawings submitted after the filing date of the application may not be used (i) to overcome any insufficiency of the specification due to lack of all enabling disclosure or otherwise inadequate disclosure therein, or (ii) to supplement the original disclosure thereof for the purpose of interpretation of the scope of any claim.

Electrical and mechanical cases usually require drawings. Chemical cases involving compounds and compositions usually do not require drawings. Chemical processes and methods cases may or may not require drawings. It is advisable to include drawings because drawings make the enabling disclosure requirement much earlier. As pointed out earlier in the chapter, when drawings are required they are considered part of the complete application and therefore necessary for assignment of a filing date and application serial number.

The Patent Office will accept "informal" drawings such as photo prints or xerographic prints on plain paper for the purpose of assigning a filing date. Sometimes in an effort to make a deadline, applications are submitted with informal drawings. When this happens, the applicant is notified, and formal drawings must be submitted within two months along with a fee.

The Patent Office will transfer drawings from a prior application to a new continuing application provided there is a written declaration of abandonment of the old application. Also, in reissue applications, the drawings will be transferred at request.

The Patent Office has rules established for the standards of the drawings. A copy of these can be obtained from the Patent Office.

FILING FEE

To complete application, the applicant must submit the required filing fee. The filing fee for a utility patent consists of a basic fee and possible additional fees. The basic fee entitles the applicant to have twenty claims examined, including not more than three independent claims. An additional fee is required for independent claims in excess of three.

During prosecution of the application, additional fees may be required. Additional fees must be paid before the amendments will be entered; these should therefore be paid with the amendments.

A copy of the fee schedule can be obtained from the Patent Office. Since the schedule changes from time to time, it is not included.

CONTINUING APPLICATIONS

Sometimes an applicant will submit one or more continuing applications following the original application. The purpose of the continuing application is to refine the application during prosecution. Due to statutory time bars, the preparer of the application may not have adequate time to prepare the best application with the original application. Continuing application practice affords an opportunity to submit a subsequent application to claim the invention in a more proper fashion.

Continuing applications are by definition applications that claim subject matter disclosed in an earlier filed application by the same inventor. The continuing application must be filed in the Patent Office before the first application has either issued or become abandoned. The major benefit of the continuing application is that it retains the benefit of the filing date of the earlier filed application.

Continuing applications are classified as continuation applications, continuation-in-part applications, and divisional applications. This section is concerned with the continuation application and the continuation-in-part application. The divisional application is a special type of continuing application discussed in Chapter 6.

Title 35, United States Code, Section 120, relates to entitlement of continuing applications to an earlier filed United States patent application and is cited below.

> An application for patent for an invention disclosed in the manner provided by the first paragraph of section 112 of this title in an application previously filed in the United States, or as provided by section 363 of this title, by the same inventor shall have the same effect, as to such invention, as though filed on the date of the prior application, if filed before the patenting or abandonment of or termination of proceedings on the first application or on an application similarly entitled to the benefit of the filing date of the first application and if it contains or is amended to contain a specific reference to the earlier filed application.

To obtain the benefit of the earlier filing date, (1) the continuing application must have been adequately disclosed in the application from which it is a continuation of, (2) the two applications must be concurrently prosecuted in the Patent Office, and (3) the two applications must be filed in the name of the same inventor.

A continuation application is one that meets all three of the above requirements, and it has the benefit of the filing date of the original application. This type of application contains only subject matter which was fully disclosed in a previous application. It contains no "new matter," that is, matter not disclosed in the original application.

A continuation-in-part application is a continuation of an original application to the extent that it discloses the same subject matter as disclosed in the original, but it also includes "new matter." New matter is any subject matter not disclosed in the original application. New matter and the claims which rely on it for enabling disclosure are not entitled to the benefit of the earlier filing date of the original application, but are entitled to the filing date on which the new matter was submitted to the Patent Office. However, any claims included in the continuation-in-part application which were fully disclosed in the original application are entitled to the filing date of the original disclosure.

Chapter 5

CLAIMS

INTRODUCTION

The claims are concise, descriptive statements at the end of the specification which point out and distinctly claim the subject matter of the invention. The purpose of the claims is to distinguish the invention from prior art and to define the parameters of the invention. From these statements, the Patent Office determines the novelty, the nonobviousness, and the utility of the subject matter during the examination of the application. When the patent issues, it is these descriptive statements that define the property rights of the patentee. Therefore properly drawn claims are written narrowly enough to distinguish the subject matter of the invention from prior art and are written broadly enough to preclude others from infringing the property rights of the inventor.

Claim drafting is the most important aspect of preparing a patent application. It is the claims that transform the invention into patent property. Well-written claims are prosecuted with ease before the Patent Office; poorly written claims are prosecuted with difficulty.

It is beyond the scope of this book to teach the craft of claim drafting. The purpose of this chapter is to introduce the format used in the claim, to define the judicial interpretation of certain terms used in the claim, and to state some of the rules of claim drafting. The objective is to familiarize the reader with the nature of claims to the extent that he can interpret claims when reading issued patents. For those not familiar with claims, it may be helpful to read the claims in the sample patent that appears in the appendix to this book.

STATUTORY BASIS

The statutory basis for the claims is found in Section 112 of the Patent Code and states in pertinent part:

The specification shall conclude with one or more claims particularly pointing out and distinctly claiming the subject matter which the applicant regards as his invention.

A claim may be written in independent or, if the nature of the case admits, in dependent or multiple dependent form.

Subject to the following paragraph, a claim in dependent form shall contain a reference to a claim previously set forth and then specify a further limitation of the subject matter claimed. A claim in dependent form shall be construed to incorporate by reference all the limitations of the claim to which it refers.

A claim in multiple dependent form shall contain a reference, in the alternative only, to more than one claim previously set forth and then specify a further limitation of the subject matter claimed. A multiple dependent claim shall not serve as a basis for any other multiple dependent claim. A multiple dependent claim shall be construed to incorporate by reference all the limitations of the particular claim in relation to which it is being considered.

An element in a claim for a combination may be expressed as a means or step for performing a specified function without the recital of structure, material, or acts in support thereof, and such claim shall be construed to cover the corresponding structure, material, or acts described in the specification and equivalents thereof.

Claims may be written in an independent form or a dependent form. An independent claim is a claim that stands alone on its own merits; it has no reference to any other claim. This type of claim defines the subject matter that it is claiming within itself. One can determine if the claim is being infringed by reading only that claim since it has no reference to any other claim.

A dependent claim is a claim that refers to another claim and places additional limitation on the subject matter of the claim to which it refers. The dependent claim can modify an element of the parent claim, or it can add an element to the parent claim. The dependent claim by its nature incorporates by reference all the limitations of the claim to which it refers. Therefore when determining if the claim is being infringed, one must consider both the dependent claim and the claim to which it refers.

A dependent claim can be dependent upon an independent claim, or it can be dependent upon another dependent claim. In addition, there can be a multiple dependency of claims. A multiple dependent claim refers back to more than one preceding independent or dependent claim. A multiple dependent claim cannot, however, serve as the basis for any other multiple dependent claim. A multiple dependent claim incorporates all the limitations of all the claims to which it refers. When determining if the claim is being infringed, one must consider the multiple dependent claim as well as all claims to which it refers.

There is sometimes a misunderstanding of the validity of a dependent claim or multiple dependent claim when claims to which these refer are judged invalid. This does not affect the validity of the dependent or multiple dependent claim. Section 282 of the Patent Code states in pertinent part "dependent or multiple dependent claims shall be presumed valid even though dependent upon an invalid claim." During litigation, each claim of the patent, whether independent, dependent, or multiple dependent, is presumed valid independently of the validity of other claims. Each is tried on its own merit.

FORMAT FOR CLAIMS

There are no statutory requirements regarding the form of claims, but, the Patent Office has over the years established certain guidelines for claim drafting. One such guideline is that each claim must be the object of a sentence, or phrase, where the sentence begins with "I claim" or with "What is claimed is." This phrase is followed by the claim which includes a preamble, a transition term, and the body.

The claim may be a single sentence, phrase, or combination of phrases which can be distinguished by semicolons, colons, and other grammatical notations, except a period. The claim must end in a period, but there can be no periods within the body of the claim. If this form is not used by the claim drafter, it will be inserted by the examiner when the patent issues. All recently issued patents are printed in this format.

As indicated above, the claim itself consists of a preamble, a transition term, and a body. Following is a discussion of the purpose of each of these.

Preamble

The preamble has no legal effect on the claim, but, it does constitute an important part of the claim. It sets forth the nature of the patentable invention, whether it is a process, machine, manufacture, or composition of matter. The preamble should distinguish the class of the invention and then further define some characteristic of the invention. The same preamble should be used in all subsequent dependent claims as is originally used in the independent claim, although it is not necessary to include the description.

Transition Terms

For independent claims, the three transitional terms most often used in the transition from the preamble to the body of the claim are "comprising," "consisting of," and "consisting essentially of." Over the years, these terms have developed a judicially recognized interpretation as they relate to patent claims. Therefore, the selection of the transition term has a significant effect on the scope of the claim.

The most commonly used transitional term is "comprising." This term is judicially interpreted to mean that the claimed invention includes all elements stated in the body of the claim, but it may also contain an unlimited number of additional, unrecited elements. The use of this term results in the so-called "open end" claim and is often used for claiming machines and articles of manufacture. When reading a claim, the term "comprising" can be replaced by "including." Any prior art which has the same elements anticipates the claim, any prior art which has the same elements plus other elements anticipates the claim. Any device that has the recited elements infringes the claim; any device that has the recited elements plus other elements infringes the claim.

More restrictive transitional phrases are the terms "consisting of" or "which consist of." These terms are judicially interpreted to mean that the claimed invention consists of only the elements stated in the body of the claim. Therefore, any embodiment that does not contain the exact (no more or no fewer) elements as those stated in the claim does not anticipate the claim. The use of this term results in the so-called "closed end" claim. Prior art must have the exact elements as recited in the claim to anticipate the claim. If the art has fewer elements, more elements, or different elements from those specified in the claim, it does not anticipate the claim. An embodiment must have exactly the same elements, no more or no fewer, as recited in the claim to infringe it.

The "closed" type claim is often used in the chemical and process arts. With composition of matter claims, the terminology does allow the presence of small amounts of chemical impurities. With process claims, the steps must be exact.

Somewhere between the "open ended" transitional phrase and the "closed ended" phrase is the term "consisting essentially of." This term is judicially interpreted to mean that the claim consists of the recited elements, but may also include other unrecited elements which do not affect the characteristics of the recited elements. This type of claim is referred to as a "partially closed" claim and is used almost exclusively in the chemical arts.

A dependent claim can either modify an element of the parent claim, or it can add an element to the parent claim. When the dependent claim modifies an element of the parent claim, the transition phrase used is "wherein" followed by the modifying language. The term "wherein" indicates that no element is being added, but that a previously recited element is being modified.

When the dependent claim adds an element to the parent claim, the transitional phrase may be "further comprising," "which further comprises," or "further consisting essentially of." These indicate that an element is being added to the parent claim. A dependent claim adding an element to a parent claim using the closed-end phrase "consisting of" would be improper.

Body

Following the transitional phrase is the body of the claim. The purpose of the body of the claim is to recite the elements of the invention, to describe how the elements are associated with each other, and to describe how the elements function together to make the invention. The elements are the atoms or groups in a chemical application, the steps in a process application, and the structural parts in a machine or article of manufacture application.

Each essential element of the inventive combination must be recited in the body of the claim, where the element is essential if it is involved in the function of the invention. Naming the element is up to the drafter; a court decision has ruled that the draftsman may be his own "lexicographer." Once the element has been named, however, it is important to be consistent in terminology.

When introducing an element into the claim, it is necessary that an indefinite article such as "a, an, one, two," etc., be used. It is not permissible to use the definite article "the." Whenever an element is referred to for a second time, however, it is permissible to use the definite article "the," or more properly the term *said*. The word *said* is used by most draftsmen in lieu of "the" and it is clearly proper so long as there is an antecedent for the element.

Following the name of the element, there should be a description of the element.

This description may describe the parts, features, or function of the element, depending upon the nature of the element. For process claims, the description might include temperature, duration of the step, etc. For compound claims, the description might include physical constants of the compound.

For mechanical and electrical cases, the last paragraph of Section 112 provides a method of defining certain elements in terms of their function by the "means plus function clause." An element in a claim for a combination can be expressed as a means or step for performing a specified function without reciting the structure, material, or acts in support of the element, and the claim is constructed to cover the corresponding structure described in the specification.

A properly written claim describes how the elements are connected together and associated with each other. When the elements are left dangling in space, it is more difficult to understand the invention. Therefore, the cited elements should always be structurally related. With machines it is good to start at the base and work upward; with articles of manufacture, a reference point is selected and the discussion proceeds from there; with processes, the steps of the process are related in their working sequence. Well written claims describe how the cited elements function together to make the invention work.

With mechanical and electrical inventions, this can be done quite readily by referring to the drawings.

SPECIAL FORM CLAIMS

Over the years, the Patent Office and courts have adapted a special form for improvement claims (Jepson claims) and chemical claims (Markush claims).

Improvement Claims

For inventions which involve an improvement of an element in a known combination of elements, the Patent Office provides a special form for the claims commonly called a Jepson-type claim after a former commissioner of patents. Independent Jepson claims contain, in the following order, a preamble consisting of a general description of all the elements or steps of the claimed combination which are conventional or known, a phrase such as "wherein the improvement comprises," and those elements, steps, and relationships which constitute the portion of the claimed combination which the applicant considers to be the improvement. The body of the claim is essentially the same as with other types of claims.

Markush Expressions

A special type of claim used exclusively in the chemical arts is the "Markush expression." This type of claim allows one to claim several chemical elements or groups at a given position when they function equivalently within a claim.

An example of a Markush expression is set forth below.

A compound of the formula

$$
\begin{array}{ccc}
H & & H \\
| & & | \\
R^1\text{-}C & - & C\text{-}R^3 \\
| & & | \\
R^2 & & H
\end{array}
$$

wherein R is methyl or ethyl, R2 is chlorine or bromine, and R3 is hydrogen or methyl.

Usually the members of the Markush group belong to a chemical class such as halogens, alkyl, etc. These expressions are proper only with chemical inventions using the closed-end transition term.

SCOPE OF CLAIMS

The purpose of the claims is to distinguish the invention over prior art during examination of the application and to define the property rights of the patentee when the patent issues. In order to obtain maximum property rights for the inventor, the patent should contain claims drawn as close to prior art as possible, that is, broad claims. The possibility exists, however, that sometime during the life of the patent, certain broad claims might be invalidated because of prior art which was not considered during examination of the application. If the patent contained only a broad claim, the patentee would have nothing. A patent therefore should contain a series of claims ranging from as broad as prior art will allow to as narrow as will be of any practical use to the patentee. Thus in the event that certain broad claims are invalidated, the patentee will still have the narrow claims remaining on his patent.

The method by which claims are made progressively narrower is by dependent claims. As previously discussed, the dependent claim adds an element to, or modifies an element of, the claim to which it refers. This makes the claim narrower as elements are added to it or previous elements are modified. The exception to this is the chemical claim, which becomes broader as elements are added.

Take for example the patent reprinted as the appendix to this book. Claim 1 is a broad independent claim setting forth a monitoring device for monitoring the data line between a fuel dispenser controller and a fuel dispenser. The next claim is a dependent claim referring to Claim 1. Since the transitional phrase "wherein" is used in the dependent claim, the claim is not adding an element to the device, but modifying an existing element. It is placing an additional restriction on the device.

Chapter 6

EXAMINATION OF
THE APPLICATION

INTRODUCTION

When a patent application first reaches the Patent Office, it is directed to the Application Division to determine if the presented documents constitute a "complete" application. A complete application includes a specification with claims, an oath or declaration, drawings when appropriate, and the required filing fee. If these are in order, a filing date and application serial number are assigned to the application. The Application Division then reads the application to determine if there are any insufficiencies in the matter of form, such as a missing signature, missing address, etc. This reading does not touch on the substantive content of the application as it relates to patentability of the subject matter, but concerns matters of form only.

If the application contains drawings, these are forwarded to the official draftsman for review. The extent of his inspection is to determine the formality of the drawings, not their substantive content. Any informalities noted by the Application Division or official draftsman are later brought to the attention of the examiner when the application is assigned to an art group for substantive examination.

Next the application is classified by the Classification Division for the purpose of directing the application to the appropriate examining group. The primary basis for classification is the utility of the invention claimed in the application. At present, there are approximately 300 subject classes and 75,000 subclasses. Different examining groups have responsibility for examining applications in different areas of technology in accordance with the classifications system.

The Patent Office has about 1,350 patent examiners. The examiners are divided up into 15 "examining groups," of which 5 groups are chemical examining groups, 5 are mechanical examining groups, and 5 are electrical examining groups. The examining groups are subdivided into "group art units," of which there are approximately 100. Each is assigned one or more general areas of technology.

An examining group is directed by a group director, and each group art unit within the examining group is headed by a supervisory primary examiner. Within the group art unit, there are a number of primary examiners who function independently and a number of assistant examiners who function under the supervisory primary examiner. Examiners are generally well trained in their area of technology by education and experience.

Flow chart 2 summarizes the sequence of events during the prosecution phase of the patent application and the various actions that the examiner may

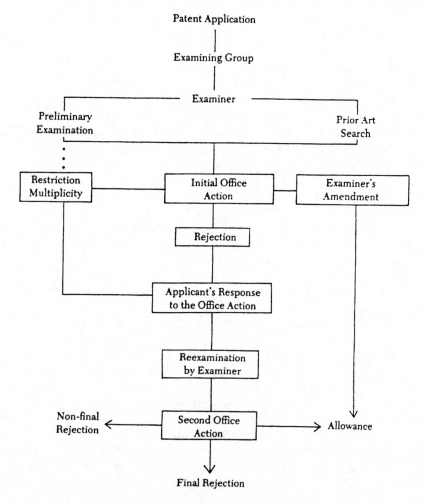

Flow chart 2. Prosecution of the Application.

take. In general, the examiner may "reject" a claim, or he may "object" to a claim. A rejection involves the substance of the claim as it relates to the statutory requirements of patentability; an objection relates the form of claim that does not conform to Patent Office rules. This distinction is important since rejections can be appealed to the Patent Board of Appeals and objections can be petitioned to the commissioner of patents.

EXAMINATION

As indicated, applications accepted as complete are assigned to the examining group having charge of the class of invention to which the application relates for examination.

Order of Examination

Once the application reaches the examiner, the Patent Office has guidelines relating to the order in which applications are taken up for initial examination. One guideline provides that applications should be taken up for initial examination by the examiner in the order of the oldest effective United States filing date. For new applications not relying on foreign priority, this is the date assigned by the application division; for continuation or division of application, this is the filing date of the parent application; for applications relying on foreign priority, this is the filing date of the first foreign-filed application.

In practice, the guideline relating to the oldest effective United States filing date is not always strictly adhered to. Many examiners tend to examine in accordance with the actual United States filing date. This is particularly true when applications depend on a foreign filing date. In addition, some group directors allow their examiners to take up a group of cases in very related subject matter, not strictly adhering to their relative filing dates. This allows the examiners greater efficiency in searching the prior art. It should be noted, however, that there is rarely a gross misuse of this guideline.

Certain applications are designated as "special" and are advanced out of the normal order for examination as established by the above guideline. Listed below is a list of special cases where the application is advanced out of turn. These cases are set forth in Section 708.01 of the *Manual of Patent Examining Procedure.*

> 1. Cases where the invention is deemed of particular importance to some branch of public service.
> 2. Cases which have been transferred from one art unit or group to another.

3. Cases pending more than five years.
4. Cases involving application for reissue.
5. Cases referred back to the examiner by court proceedings.
6. Cases where it appears that the application will be involved in an interference proceeding.
7. Cases ready for issue.
8. Cases in condition for final rejection.
9. Cases made special as a result of a petition from the applicant.

In all the cases above except example 9 (cases made special as a result of a petition from the applicant), cases are made special by internal procedure of the Patent Office. In example 9, the applicant submits a petition to have his application made special. This is accomplished by filing a petition to make his application special and is discussed later in this chapter.

During the prosecution of the application, there is various correspondence between the examiner and the applicant. The applicant responds to this correspondence by submitting "responses" such as amendments, affidavits, petitions, etc., which require further action by the examiner. It is Patent Office procedure for the examiner to act on these responses according to the date on which the applicant has taken his last action; the guideline is that preference is given to older cases.

Examiner's Preliminary Examination and Search

When the examiner takes up a new case, his first effort is to examine the application for informalities. He determines if the form of the application meets the guidelines established by the Patent Office as discussed in Chapter 4. He also at this time overviews certain technological aspects of the application. For example, he determines if the technology in the application is relevant to the art group to which the application has been assigned. He also determines if the drawings are sufficient to meet the requirements of the drawings as established by the Patent Office.

The purpose of the preliminary examination is to determine if the application is in proper form to warrant a substantive examination. If not, the applicant is so notified and is given a 30-day time period to correct the informalities.

During the preliminary examination, the examiner may conclude that the application contains claims directed to more than one independent and distinct invention. If so, he will require that the applicant select claims directed to a single invention (restriction requirement) as discussed later in this chapter. The examiner may at this time also allege that there is a multiplicity of claims, that is, an unreasonable number of claims in view of the nature and scope of the application. In this case the examiner will request that the applicant select a specified number of claims for the purpose of initial examination as also discussed later in this chapter. The above requirements are made before the prior art search.

If the application is in condition for a substantive examination, the next step is the examiner's prior art search. Patent Office Rule 104 relates to the examiner's prior art search and is cited below in pertinent part.

> (a) On taking up an application for examination, the examiner shall make a thorough investigation of the available prior art relating to the subject matter of the invention sought to be patented. The examination shall be complete with respect both to compliance of the application with the statutes and rules to the patentability of the invention as claimed, as well as with respect to matters of form, unless otherwise indicated.

The examiner searches the United States patents in the classified art collection assigned to him along with literature references and foreign patents in that collection. The resulting search is fairly complete with regard to issued United States patents, but, it may be less than complete with regard to foreign patents and literature. The examiners' art collection contains only limited information on foreign patents and literature.

Considering the case load of examiners, they usually do a commendable job in this search. It should never be assumed, however, that this is a complete search. Examiners clearly do not have the time or the resources for a complete search. If the validity issue arises, it can be expected that additional pertinent prior art will be presented by the party asserting invalidity. Such art may come from foreign patents, foreign publications, subclasses not in the examiners' collection, etc. The party asserting invalidity does a much more comprehensive search than is possible by an examiner.

The examiner keeps a record on the file wrapper of the application of his field of search. This record is very important if the patentee wants to do additional searching because this record tells the patentee where the examiner did not search. Also, this forms the prior art record which may be of importance if the validity question is raised as discussed in the chapter on infringement. Judges rarely overrule an examiner's opinion on the pertinence of a prior art reference.

INITIAL OFFICE ACTION

Following the examiner's prior art search, the examiner makes a decision on the case and thereafter notifies the applicant of this decision. This forms the initial office action and is the examiner's first letter with regard to the patent application. The statutory requirement for this action is found in Section 132 of the Patent Code cited below.

> Wherever, on examination, any claim for a patent is rejected, or any objection
> or requirement made, the Commissioner shall notify the applicant thereof,
> stating the reasons for such rejection, or objection or requirement, together with
> such information and references as may be useful in judging of the propriety
> of continuing the prosecution of his application; and if after receiving such
> notice, the applicant persists in his claim for a patent, with or without amend-
> ment, the application shall be reexamined. No amendment shall introduce new
> matter into the disclosure of the invention.

This action must eventually be in a written form. However, the examiner
sometimes first telephones the patent practitioner to discuss his decision, and
the substance of the phone call is followed by a written account.

By this statute the examiner must state the reasons for the rejection, objection, or requirement for restriction, along with references on which these are
based. For prior art rejections, it is office policy to include one copy of each U.S.
patent reference cited. When a foreign patent or printed publication is cited,
sufficient information is given to allow location of these by the applicant.

Along with this, the examiner must indicate a time period for response
by the applicant as set forth in Section 133 of the Patent Code cited below.

> Upon failure of the applicant to prosecute the application within six months
> after any action therein, of which notice has been given or mailed to the appli-
> cant, or within such shorter time, not less than thirty days, as fixed by the Com-
> missioner in such action, the application shall be regarded as abandoned by the
> parties thereto, unless it can be shown to the satisfaction of the Commissioner
> that such delay was unavoidable.

If a shorter time period for response is not set by the examiner, the applicant has six full months to respond. Usually a shorter period of three months
is set by the examiner. If the applicant does not respond during this period,
the application becomes abandoned. The period of response is based on the
date on which the action was mailed.

The initial office action may be on the merits of the case which relate to
the statutory requirements for patentability resulting from the examiner's
substantive examination, or it may not be on merits where the action is taken
before the examiner's substantive examination. Actions on merit include
allowance of claims, examiner's amendments to claims, or more commonly, rejections of claims. Initial office actions not on the merits include restriction requirements and actions relating to undue multiplicity. The latter two actions
are set forth before the examiner's prior art search, and do not therefore consider the statutory requirements for patentability.

Requirement for Restriction

If when examining a case, the examiner decides that the application contains claims directed to more than one independent and distinct invention, he

may require the applicant to select claims directed to a single invention. This constitutes an objection, the authority of which is found in Section 121 of the Patent Code cited below.

> If two or more independent and distinct inventions are claimed in one application, the Commissioner may require the application to be restricted to one of the inventions. If the other invention is made the subject of a divisional application which complies with the requirements of section 120 of this title it shall be entitled to the benefit of the filing date of the original application. A patent issuing on an application with respect to which a requirement for restriction under this section has been made, or on an application filed as a result of such a requirement, shall not be used as a reference either in the Patent and Trademark Office or in the courts against a divisional application or against the original application or any patent issued on either of them, if the divisional application is filed before the issuance of the patent on the other application. If a divisional application is directed solely to subject matter described and claimed in the original application as filed, the Commissioner may dispense with signing and execution by the inventor. The validity of a patent shall not be questioned for failure of commissioner to require the application to be restricted to one invention.

With a restriction requirement, the examiner informs the applicant which claims are directed to which distinct invention, and he requests the applicant to select a single disclosed species for the purpose of initial examination. After selection, the examiner begins a substantive examination of the elected species, and the nonelected claims are withdrawn from prosecution. If the applicant so desires, the nonelected invention can be filed in a divisional application. When a divisional application is filed on the nonelected invention, the application is entitled to the filing date of the original application. In addition, the divisional application cannot be used as a prior art reference against the original application, nor can the original application be used as prior art against the divisional application.

It is common practice for the examiner to telephone the practitioner of record to ask if he will make an oral selection, with or without traverse. If an oral selection is made, the examiner proceeds to examine the selected species. In this case, the initial action includes the record of the restriction requirement and the result of the examination on merits of the selected claims.

If the practitioner refuses to make an oral selection, the initial action relates only to the restriction requirement. The applicant has thirty days to respond by making selection or the application becomes abandoned. If the selection was traversed in either the oral or written selection, the requirement may be petitioned.

Undue Multiplicity

Title 37, Code of Federal Regulations, Section 1.75(b), relates to the undue multiplicity rejection and is cited below.

(b) More than one claim may be presented provided they differ substantially from each other and are not unduly multiplied. If the examiner feels that an unreasonable number of claims in view of the scope of invention has been included in the application, he may reject the application on the grounds of multiplicity. In such cases, the examiner will telephone the practitioner prosecuting the case to explain the rejection, to indicate the number of claims considered permissible by the examiner, and to request selection. If selection is made, the examiner proceeds to examine the selected claims on merit. In this case, the Initial Action consists of a record of the multiplicity rejection and the results of the examination on merits of the selected claims.

If the applicant refuses to select, the initial action relates only to the multiplicity rejection, which rejects all the total number of claims and then specifies what number of claims would be considered permissible. If the applicant selects claims without exceeding the number specified by the examiner, these are examined on merit. A rejection on undue multiplicity is a statutory rejection and is therefore reviewable by the Patent Office Board of Appeals if the applicant traverses the rejection.

First Action Allowance

Occasionally after examination, the examiner finds a case in condition for allowance as filed except for minor formal matters. When this happens, the initial office action is a first action allowance. Very few applications are allowed as filed, however.

Examiner's Amendment

In other cases, the examiner finds all claims to be allowable except for minor insufficiencies which can be corrected by amendment. The examiner obtains oral consent from the applicant and proceeds to make the necessary amendments to either specification or claims. These amendments are filed, and the application subsequently allowed. Again, this is rare.

Rejections on Merits

If after substantive examination, the invention is not considered to be patentable or not considered patentable as claimed, the claims are rejected in part or total. The initial office action in a vast majority of the cases involves rejection of all or part of the submitted claims.

When the examiner finds that the invention as defined by the claims does not meet the conditions for novelty as set forth in Section 102, the claims are rejected as being "anticipated" by prior art. When the examiner finds the invention as defined by the claims is not sufficiently different from prior art under Section 103, the claims are rejected as being "obvious over" prior art. The following section discusses these rejections.

REJECTION OF CLAIMS

A rejection of a claim involves the substance of the claim as it relates to the statutory requirements for patentability. A great majority of all rejections are based on Section 102, which sets forth the novelty requirement for patentability, and Section 103, which sets forth the nonobvious requirement for patentability. Most rejections grounded on these sections are based on prior art, the exceptions being subsection (c) of Section 102, which involves abandonment of the invention, and subsection (f) rejections, which involve inventorship.

Other statutory rejections sometimes encountered are rejections grounded on Sections 101, which involve nonstatutory subject matter, lack of utility, and double patenting, and Section 112 rejections, which involve insufficient disclosure of the invention. These rejections are not based on prior art, but on the failure of the invention or the patent application to meet the requirements in the respective statute.

The examiner must state the reasons for the rejection along with such information and references as may be useful to the applicant for continuing the prosecution of his application. It is Patent Office policy to supply the applicant with a copy of each United States patent cited in the rejection and sufficient information to allow the applicant to locate any foreign patent or literature reference cited.

Based on Section 102

For a Section 102 rejection to be proper, the prior art reference must both describe and illustrate the subject matter of the claim "within four corners"; the reference must clearly anticipate the claim. Most courts have interpreted that a Section 102 rejection must meet this requirement, and this is the interpretation followed by the Patent Office. A Section 102 rejection usually involves a single reference. If a combination of references are used, this is most often an improper rejection.

As indicated above, a reference cited in a Section 102 rejection must both describe and illustrate the subject matter of the application. The reference may do this in varying degrees, however. At one extreme, the reference may disclose the entire concept of the invention as well as the structure of the invention, thereby anticipating all claims. This rejection can be overcome only by swearing behind the reference date by a Rule 131 affidavit as discussed in the next chapter.

On the other hand, the reference may only anticipate one or two of the claims of the application and not disclose the entire inventive concept. In this case, it is a matter of redrafting the claims by amendment to overcome the reference. This situation is quite common since claims are often submitted which are drafted as broad as prior art will permit, and the broader claims are rejected whereas the narrower claims are allowed.

Based on Section 103

When the examiner is unable to find a reference that identically discloses or describes the subject matter as required for a Section 102 rejection, he may find one or more references which make the subject matter obvious to a person skilled in the relevant art. Section 103 states that the subject matter must be nonobvious to a person skilled in the relevant art at the time of the invention for a valid patent to issue. For a Section 103 rejection, the examiner often uses a combination of two or more references. Three terms utilized by the examiner in combining prior art references are "in view of," "and," and "or."

When making a Section 103 rejection, the examiner must state the differences in the claimed subject matter and the cited references, state the proposed modifications of the cited references required to arrive at the claimed subject matter, and explain why such modifications would be obvious to a person skilled in the art at the time of the invention. This requirement is set forth in the *Manual of Patent Examining Procedure,* Section 706.02.

Based on Section 101

Rejections based on Section 101 of the Patent Code are not based on prior art, but on a failure to meet the requirements of this statute. Possible Section 101 rejections are nonstatutory subject matter, lack of utility, and double patenting. Section 101 is cited below.

> Whoever invents or discovers any new and useful process, machine, manufacture, or composition of matter, or any new and useful improvement thereof, may obtain a patent therefor, subject to the conditions and requirements of this title.

NONSTATUTORY SUBJECT MATTER. By this statute the invention or discovery must be a process, machine, manufacture, or composition of matter to be patentable. Court interpretation of this statute has held certain subject matter to be unpatentable, including naturally occurring articles, printed matter, methods of doing business, scientific principles, metal steps, and aggregations.

Naturally occurring articles belong to the public and therefore cannot be taken away. However, if a naturally occurring article is changed, modified in some way, or used for a purpose other than that in nature, a Section 101 rejection would not properly apply.

The arrangement of printed matter cannot be patented, but it can be copyrighted. Court interpretation of this statute does not, however, relate to machines or processes used in the printing art; these are patentable.

Methods of doing business do not meet the patentability requirements of Section 101. The purpose of the patenting system is to promote progress in the arts; court interpretation of this statute has held that methods of doing business do not fulfill this requirement.

Scientific principles and laws of nature as a general rule are not patentable. This precludes the patenting of mathematical equations, etc.

As a general rule, mental steps cannot be patented. This issue has played a significant role in patenting computer-type processes, particularly software. Since many of these steps can conceivably be carried out in the mind, there is a thin line between what is properly rejected and allowed with this subject matter.

An aggregation is not an invention and therefore not patentable under Section 101. With an aggregation there is no cooperation between the elements of the claim and therefore not an invention by definition.

UTILITY. The language of Section 101 makes it clear that any process, machine, manufacture, or composition of matter, or any improvement thereof, must have utility to be patentable. During examination of the application, the examiner accesses utility of the invention. For machines and articles of manufacture, utility is usually self-evident from the disclosure in the application. Likewise, the utility of most processes is self-evident from the ability of the process to produce its stated objective.

The utility of chemical compounds and compositions is often more difficult to establish and is discussed in Chapter 3. Suffice it to say at this point that the Patent Office requires that compounds or compositions have practical utility in the chemical arts to meet the utility requirement. Often the applicant must submit affidavits to support utility.

DOUBLE PATENTING. The use of the term "a patent" in Section 101 has been interpreted to limit an invention to one patent. Therefore, a double patenting rejection can occur when the inventor files more than one application claiming the same invention. This is a statutory rejection, and it involves claims directed to exactly the same invention.

When two applications by the same inventor claim subject matter which is not identical, but sufficiently close that the invention claimed in one application would be obvious in view of the invention claimed in the other application, the second application may be rejected by an "obviousness-type" double patenting rejection. This is a judicially established rejection. In applying this type of rejection, it is proper for the examiner to use prior art references such as patents and printed publications to determine whether the claims in the second application would be obvious to one skilled in the art in view of the reference. A double patenting rejection can most often be overcome with a "terminal disclaimer." The terminal disclaimer provides that a patent granted on an application filed with a terminal disclaimer can be enforceable only for the same time that the commonly owned patent forming the rejection is enforceable.

Based on Section 112

Section 112 of the Patent Code requires that the application contain an enabling disclosure of the invention. The application must contain a written description of the invention, it must describe the manner and process of making and using the invention, and it must set forth the best mode contemplated by the inventor for carrying out his invention at the time the application was filed. Failure to meet these requirements results in a Section 112 rejection. The first paragraph of Section 112 relates to this type of rejection and is cited below in pertinent part.

> The specification shall contain a written description of the invention, and of the manner and process of making and using it, in such full, clear, concise, and exact terms as to enable any person skilled in the art to which it pertains, or with which it is most nearly connected, to make and use the same,and shall set forth the best mode contemplated by the inventor of carrying out his invention.

The disclosure requirement can rely only on the disclosure set forth in the application as originally filed in the Patent Office. The specification cannot be amended to meet this requirement during the prosecution phase. The purpose of the enabling disclosure is to teach the invention to the public.

Failure to describe the invention can result in a Section 112 rejection. Any subject matter claimed in the application must be fully described in the specification, and it must be described in such terms that a person skilled in the art could practice the invention after reading the description without experimentation. The terms and phrases used in the claims must have clear antecedent basis in the description. The meaning of the terms in the claims must be understood by reference to the description in the specification; otherwise, a Section 112 rejection will arise.

Failure to teach the invention can result in a Section 112 rejection because the purpose of the enabling disclosure is to teach the invention to the public. Therefore, there must be a written description of the manner and process of making and using the invention. With mechanical and electrical cases, the drawings are very helpful in teaching the invention. Since chemical cases usually do not include drawings, there must be a very carefully written description in specification describing how to make and use the compound or composition.

Failure to set forth the best mode can also result in a Section 112 rejection. The applicant must set forth the best mode for carrying out his invention known to him at the time of filing the application. This is to prevent the inventor from withholding information about his invention from the public while at the same time having patent rights. Since the examiner is not in the position to determine if the best mode has been set forth, rejections on these grounds are extremely rare. However, if interparty proceedings evolve, a patent can be invalidated on these grounds.

APPLICANT'S RESPONSE AND REEXAMINATION

After receiving the initial office action on merits, the applicant must file a response within the time set in the action or the application becomes abandoned.

The applicant's response is fully described in the next chapter. Suffice it to say at this point that when a proper response is filed, the Patent Office must reexamine the application. After reexamination, the applicant is notified whether the amended claims are allowed, rejected, or objected to in the same manner as after the first examination. This forms the second office action.

Usually the second office action is made the final office action. The final office action serves to terminate prosecution of the application on merits. This action indicates claims allowable, indicates claims finally rejected and the grounds for the rejections, and formally indicates that prosecution on merits is closed.

A final office action is proper on the second or any subsequent action on merit unless the examiner introduces new grounds of rejection which cannot be overcome by amendment by the applicant whether or not the prior art is of record or he makes a rejection grounded on newly cited prior art of any claim not amended by the applicant.

Following a final rejection, the applicant has three courses of action available to him if he persists in further prosecuting the application. He may file an amendment under 37 CFR 1.116 in an effort to eliminate the grounds of the rejection as discussed in Chapter 7; he may file a continuing application under 37 CFR 1. 60 followed by abandonment of the rejected application as discussed in Chapter 4; or he may appeal the rejection under 37 CFR 1.191.

FILING DATE OF THE APPLICATION

As previously discussed, a filing date is assigned to an application on the date on which a complete and acceptable application is received at the Patent

Office. This is the "actual" United States filing date. There are, however, certain circumstances under which an application is given credit for a filing date earlier than the actual United States filing date of the application. This is the "effective" filing date. An earlier effective filing date may be based upon the following circumstances.

CONTINUING APPLICATION. With a continuing application, credit is given for an earlier filed United States patent application or an international application designating the United States for the same invention.

DIVISIONAL APPLICATION. With a divisional application, the application filed on the nonelected species is entitled to the filing date of the parent application.

FOREIGN PRIORITY. By a treaty with 78 other countries, credit is given to an earlier filed foreign or international patent application for the same invention, discussed in Chapter 12.

APPLICANT'S PETITION TO MAKE SPECIAL

In an earlier section in this chapter, certain situations were discussed in which applications are designated as "special" and advanced out of the normal order for examination. In most cases applications are made special by an internal procedure of the Patent Office. However, under certain circumstances the applicant may wish to have his application examined at an accelerated rate. In these cases the Patent Office has a procedure whereby the applicant can request special status. Special status is initiated by the applicant submitting a "petition to make special," and if this petition is granted, the application is advanced out of its normal order for examination. The petition may be submitted prior to the initial examination or any time thereafter.

The following situations are set forth in Section 708.02 of the *Manual of Patent Examining Procedure* as grounds for special status by the applicant. These include prospective manufacture, infringement, applicant's health and age, environmental quality, energy, inventions relating to recombinant DNA, and accelerated procedure for new applications. These grounds are summarized below along with any additional information which the applicant must submit with the petition.

MANUFACTURE. An application may be made special on the grounds of prospective manufacture. When submitting a petition on these grounds, the applicant or assignee must in addition submit a declaration including the following: a statement that the prospective manufacturer has sufficiently available capital and facilities to manufacture the invention; a statement that the prospective manufacturer will not manufacture or will not increase his present manufacture unless the patent is granted; and a statement obligating the applicant to manufacture the invention if the patent is granted. In addition to the above information from the applicant, the patent practitioner of record must file an affidavit stating that he made, or caused to be made, a prior art search, and he believes all claims of the application to be allowable.

INFRINGEMENT. If the subject matter of an application is being infringed, it may be made special on the grounds of infringement. When submitting the petition, the applicant must submit an affidavit stating that there is actually an infringing device on the market or product in use and when the alleged infringing device was first discovered. The patent practitioner of record must also submit an affidavit stating that he has compared the infringing device to the claims of the application and in his opinion the claims are being infringed. He must indicate that a prior search has been made and that in his opinion the claims of the application are allowable.

APPLICANT'S HEALTH. If the applicant's health is such that he might not be available to assist in the prosecution of his application if it ran its normal course, the application may be made special by a showing of a doctor's certificate to this effect.

APPLICANT'S AGE. If the applicant is 65 years of age or more, the application may be made special by a showing of age through a birth certificate.

ENVIRONMENTAL QUALITY. If the invention materially enhances the quality of the environment by contributing to the restoration or maintenance of air, water, or soil, the application may be made special by an affidavit explaining how the application accomplishes this result.

ENERGY. If the subject matter of the application contributes to the discovery or development of energy resources, or to more efficient utilization and conservation of energy resources, the application may be made special by an affidavit showing how the invention accomplishes this result.

INVENTIONS RELATING TO RECOMBINANT DNA. During recent years, there has been a drastic increase in the amount of research done with recombinant DNA. While many people believe this type of research may eventually be beneficial to mankind, there is concern over potential dangers when this research is done in many laboratories. With this in mind, the Patent Office may make special those applications that relate to safety features concerned with research in this field. When petitioning on these grounds, the applicant must also submit an affidavit explaining how the subject matter of the application relates to safety of research recombinant DNA.

ACCELERATED EXAMINATION FOR NEW APPLICATION. Any new application, regardless of the subject matter, may be granted special status if the applicant so petitions. A new application is defined by the Patent Office as one that has not received any examination by the examiner. Therefore, if the applicant wishes to have any new application examined at an accelerated rate, he may do so provided the petition is submitted before the examiner acts on the application.

As summarized from Section 708.02 MPEP, a new application may be granted special status provided the applicant follows these steps:

1. Submits a petition to make the application special. Since only new applications may be granted special status, the petition must be filed with application or shortly thereafter.

2. Presents claims directed to a single invention. If the Patent Office deems that all claims are not directed to a single invention, the applicant will make selection without traverse; otherwise, the special status is removed. (Divisional applications directed to the nonselected species are not made special unless a separate petition is filed relating to the divisional application.)

3. Submits a statement that a prior art search was made, specifies who made the prior art search, and presents a list of the field of search covered by the searcher such as class, subclass, publications, foreign patents, etc.

4. Submits a copy of each reference deemed pertinent to the subject matter of the claims.

5. Submits a discussion of the reference pointing out how the claimed subject matter distinguishes from the reference. When a reference is to be removed by a Rule 131 affidavit (see next chapter), the affidavit must be submitted before the application is taken up for examination, and in no event later than one month after the petition is submitted.

DEFENSIVE PUBLICATIONS

A number of companies file patent applications strictly for defensive purposes. This is done to preclude competitors from obtaining patent protection in a certain area of development; the issued patent serves as prior art against the competitor's patent application. While this approach can be effective in keeping an area of development open, it does have disadvantages. First, it is expensive, and second, it ties up patent examiners in examining applications for which no commercial development is likely.

Being aware of the above approach and problems associated with it, the Patent Office maintains a Defensive Publication Program. If an applicant agrees to waive his rights to an enforceable patent on a submitted application, the Patent Office will publish an abstract on the disclosure of the application and make the application publicly available as prior art. Patent Office Rule 139 relates to the Defensive Publication Program and is cited below.

> An applicant may waive his rights to an enforceable patent based on a pending patent application by filing in the Patent and Trademark Office a written waiver of patent rights, a consent to the publication of an abstract, and authorization to open the complete application to inspection by the general public, and a declaration of abandonment, signed by the applicant and the assignee of record or by the attorney or agent of record.

To obtain a defensive publication, the applicant must waive patent rights for that application, consent to publication of an abstract of his disclosure in the publication, and authorize the commissioner to permit inspection of the complete application by the public. The request for the defensive publication must be filed while the pending application is awaiting the initial office action or within eight months of the earliest effective United States filing date if the initial office action has issued.

There are two other points related to the defensive publication which are not set forth in Rule 139 but should be stressed. First, the applicant maintains the right to file a continuing application on the disclosure with the benefit of the original filing date if the continuing application is filed within thirty months. Second, the applicant maintains the right to participate in interference for a period of five years after the filing date of the application. It is these two rights that make the defensive publication more attractive to most companies compared to publication of the subject matter in a technical journal for defensive purposes.

When the examiner receives a request for a defensive publication, he scans the disclosure of the application to determine if it is suitable for publication. If the disclosure is clearly anticipated by prior art, lacks utility, is frivolous, etc., it will not be published.

The publication itself is an abstract of up to 200 words and a selected drawing of the disclosure in the application. These are published in the publication and kept on file in the public search room of the Patent Office. The applications are filed in the record room and are open to the public on request.

Chapter 7

APPLICANT'S RESPONSE

INTRODUCTION

If after any adverse office action the applicant persists in prosecuting his application, he must respond to the office action within the time allowed or the application becomes abandoned. He may request reexamination or reconsideration of his application, with or without amendment.

In the office action, the examiner must state the reasons for rejections, objections, or restriction requirements. He must supply information and references which might be useful to the applicant in continuing the prosecution of his application. If the rejection is based on prior art, it is Patent Office policy to include a copy of each United States patent reference cited and sufficient information to allow location of any foreign patent or printed publication cited.

After considering the office action, the applicant may file a response requesting reconsideration without making amendment, he may file a response making amendment and requesting reexamination, or both. An amendment is a response that makes a change in the application. The most commonly submitted amendment is one that changes the claim(s) to overcome a prior art rejection. Any amendment must be fully supported by the original disclosure because no new matter can be introduced into the disclosure of the invention.

In place of an amendment, it may be more appropriate for the applicant to submit an affidavit under Patent Office Rules 131 and 132 and request reconsideration. This affidavit is used to point out certain facts to the examiner which are not evident in the application or in a prior art reference. An affidavit under Rule 131 allows one to remove certain references as prior art. An affidavit under Rule 132 allows the applicant to traverse a prior art reference, that is, to support a particular interpretation of the reference which the applicant is trying to promote. A Rule 132 affidavit is used most often in response to an obviousness rejection.

Whether one can best respond by amendment, affidavit, or a combination depends on the rejection. Consider, for example, a Section 102 rejection on prior art. If the cited reference discloses the entire concept of the invention as well as the structure of the invention, the rejection can be overcome only by swearing behind the reference date of the prior art by a Rule 131 affidavit. If the reference only anticipates one or two of the claims and does not disclose the inventive concept, the rejection might be overcome by amending the rejected claims.

This chapter is concerned with the applicant's response and, in particular, the requirements of the response, the freedom that one has in making amendment to his application, and the appropriate use of affidavit to overcome rejections.

REQUIREMENTS OF THE RESPONSE

The Patent Office has established requirements for the applicant's response. These requirements must be met whether one is responding by amendment, affidavit, or both. These requirements are set forth in 37 1.111 cited below.

> (a) After the office action, if adverse in any respect, the applicant, if he persists in his application for a patent, must reply thereto and may request reexamination or reconsideration, with or without amendment.
>
> (b) In order to be entitled to reexamination or reconsideration, the applicant must make request therefor in writing, and he must distinctly and specifically point out the supposed errors in the examiner's action; the applicant must respond to every ground of objection and rejection in the prior office action (except that request may be made that objections or requirements as to form not necessary to further consideration of the claims be held in abeyance until allowable subject matter is indicated), and the applicant's action must appear throughout to be a bona fide attempt to advance the case to final action. A general allegation that the claims define a patentable invention without specifically pointing out how the language of the claims patentably distinguishes them from the references does not comply with the requirements of this section.
>
> (c) In amending an application in response to a rejection, the applicant must clearly point out the patentable novelty which he thinks the claims present in view of the state of the art disclosed by the references cited or the objections made. He must also show how the amendments avoid such references or objections.

The response must be made in writing, and it must be fully responsive to every ground of objection or rejection raised by the examiner in the office action. It must be a good faith attempt to advance the case to a final action; the mere allegation that the examiner is wrong in his actions is not a proper reason for reconsideration or reexamination. When amending the application, the applicant must point out how the amendments avoid any prior art references cited or objections made.

When a response is submitted which is not fully responsive, the examiner may deem the defect to be unintentional and allow the applicant to correct the response, provided sufficient time remains in the time set for response. If the examiner deems the defect to be intentional, he will hold the application to be abandoned.

AMENDMENTS

An amendment is a response that makes a change in the application and may be either in the specification, the claims, or the drawings. The freedom that one has in making amendment to his application depends upon the point of the examination procedure when the amendment is submitted. Patent Office Rule 115 relates to amendments in general and is cited below.

> The applicant may amend before or after the first examination and action, and also after the second or subsequent examination or reconsideration as specified in sect. 1.112 or when and as specifically required by the examiner.

According to this rule, the applicant may freely amend before or after the initial office action. After the second office action, however, there are numerous restrictions on the right to amend.

An amendment may be submitted concurrently with the application or any time after filing it. If the amendment is filed by the examiner before the initial office action, it is called a "preliminary amendment." A preliminary amendment does not have the status of original disclosure. It is therefore not considered part of the original disclosure of the application and is not considered when determining if the application meets the enabling disclosure requirement as set forth in Section 112 of the Patent Code.

By far the most common reply to the initial office action involves amending claims to conform to prior art that the examiner has cited as grounds for rejection of claims. Patent Office Rule 119 relates to amendment of claims and is cited below.

> The claims may be amended by canceling particular claims, by presenting new claims, or by rewriting particular claims as indicated in sect. 1.121. The requirements of sect. 1.111 must be complied with by pointing out the specific distinctions believed to render the claims patentable over the references in presenting arguments in support of new claims and amendments.

When responding to the initial office action on merits, the applicant has great freedom in amending the claims. The only requirement is that the claim have supporting disclosure in the originally filed application. The claims may be amended by canceling particular claims, by amending the language of particular claims, or by presenting new claims. When new or amended claims are presented, the response must point out how the new or amended claims avoid any reference or ground of rejection of record which may be pertinent.

At this point in prosecution, the applicant has a lot of freedom in amending claims. Following the second office action, the applicant is greatly restricted in his freedom to amend claims. For this reason, it is advisable at this point, when responding to the initial office action, to submit a range of new or amended claims as broad as the prior art of record will allow. This is the last point during prosecution that one has this freedom to amend.

After the response from applicant has been received, the application is reexamined or reconsidered. The applicant is thereafter notified if claims are rejected or objected to in the same manner as after the first examination. The applicant then responds to this office action in the same manner with or without amendment, but any amendments after the second office action are restricted to the rejection or to the objections or requirements made. The application is again considered and so on repeatedly until the examiner indicates that the action is final.

Second office actions are usually made final although one does occasionally receive a second nonfinal office action. This would be appropriate when the examiner finds a better prior art than that considered in the initial office action, or when the examiner significantly shifts his basis of rejection. The applicant is now more restricted in his freedom for amendment. Amendments are restricted to rejections, objections, and restriction requirements made in the second office action.

When an applicant receives a final office action, his options are limited to appealing rejected claims to the Patent Office Board of Appeals, filing a continuing application followed by abandonment of the rejected application, or amending his claims in accordance with Rule 116.

Appeal and continuing applications are discussed in other chapters. This section is concerned with Patent Office Rule 116 which relates to amendments after final action and is cited below. After final rejection, amendments can be made canceling claims or complying with any requirement of form which has been made, and amendments presenting rejected claims in better form for consideration on appeal may be admitted.

MANNER OF MAKING AMENDMENTS

When the applicant submits amendments to the Patent Office, he must set forth proper instructions in his response for the Patent Office to enter the amendments into the application. Patent Office Rule 121 relates to the manner for making amendments in the specification and claims and is cited below.

(a) Erasures, additions, insertions, or alterations of the Office file of papers and records must not be physically entered by the applicant. Amendments to the application (excluding the claims) are made by filing a paper directing or requesting that specified amendments be made. The exact word or words to be stricken out or inserted by said amendment must be specified and the precise point indicated where the deletion or insertion is to be made.

(b) Except as otherwise provided herein, a particular claim may be amended only by directions to cancel or by rewriting such claim with underlining below the word or words added and brackets around the word or words deleted. The rewriting of a claim in this form will be construed as directing the cancellation of the original claim; however, the original claim number followed by the parenthetical word "amended" must be used for the rewritten claim. If a previously rewritten claim is rewritten, underlining and bracketing will be applied in reference to the previously rewritten claim with the parenthetical expression "twice amended, three times amended," etc., following the original claim number.

(c) A particular claim may be amended in the manner indicated for the application in paragraph (a) of this section to the extent of corrections in spelling, punctuation, and typographical errors. Additional amendments in this manner will be admitted provided the changes are limited to (1) deletions and/or (2) the addition of no more than five words in any one claim. Any amendment submitted with instructions to amend particular claims but failing to conform to the provisions of paragraphs (b) and (c) of this section may be considered nonresponsive and treated accordingly.

(d) Where underlining or brackets are intended to appear in the printed patent or are properly part of the claimed material and not intended as symbolic of changes in the particular claim, amendment by rewriting in accordance with paragraph (b) of this section shall be prohibited.

(e) In reissue applications, both the descriptive portion and the claims are to be amended as specified in paragraph (a) of this section.

Subsection (b) of Rule 121 relates to amending claims. The entire claim should be rewritten in the response, the material to be added underlined, the material to be deleted placed in square brackets. From these simple instructions the Patent Office can insert or delete words in the claim. In addition, the term "amended" or "once amended," etc., should follow the original claim number.

The original numbering of the claims must be preserved throughout the prosecution. When claims are canceled, the remaining claims must not be renumbered. When claims are added, they must be numbered by the applicant consecutively beginning with the number next following the highest numbered claim previously presented. When the application is ready for allowance, the examiner renumbers the claims consecutively in the order in which they appear.

Subsection (a) of Patent Office Rule 121 relates to amending the specification. By this rule it is necessary to set forth exactly where the insertion or deletion is to be made. For example, on page 6, line 3, rewrite "chemical" as --chemical--. While not included in this rule, the use of quotation marks for deletion and double hyphens for insertion is customary.

No change in the drawing may be made except by permission of the Patent Office. Permissible changes in the construction shown in any drawing may be made only by the office. A sketch in permanent ink showing proposed changes, to become part of the record, must be filed. The paper requesting amendments to the drawing should be separate from other papers. Substitute drawings are not ordinarily admitted in any case unless required by the office.

AFFIDAVITS

An affidavit, or declaration, is another type of response commonly submitted to the Patent Office in response to an office action. The function of the affidavit is to point out certain facts to the examiner which are not evident in either the application or in a cited prior art reference. The terms affidavit and declaration are used interchangeably in this text; the Patent Office accepts either. They differ only in that the affidavit has been notarized, whereas the declaration contains a warning to the declarant.

Affidavits are submitted under either Patent Office Rule 131 or 132. Rule 131 allows one to remove certain cited prior art as reference; Rule 132 allows one to submit relevant evidence in the case for the purpose of stressing a point.

As discussed in Chapter 2, there are certain potential rights associated with the date of invention in United States patent law. The Patent Office assumes that a reference is prior art if the effective date of reference is before the effective filing date of the application. This has to be the case since the effective filing date of the application is the only date of invention of which the Patent Office has knowledge. If no statutory time bar exists, however, the reference may be removed as prior art if the applicant can demonstrate relevant acts of invention in this country before the effective date of the reference. Patent Office Rule 131 provides for affidavits for this purpose and is cited below.

(a) When any claim of an application is rejected on reference to a domestic patent which substantially shows or describes but does not claim the rejected invention, or on reference to a foreign patent or to a printed publication, and the applicant shall make oath or declaration as to facts showing a completion of the invention in this country before the filing date of the application on which the domestic patent issued, or before the date of the foreign patent, or before the date of the printed publication, then the patent or publication cited shall not bar the grant of a patent to the applicant, unless the date of such patent or printed publication be more than one year prior to the date on which the application was filed in this country.

(b) The showing of facts shall be such, in character and weight, as to establish reduction to practice prior to the effective date of the reference, or conception of the invention prior to the effective date of the reference coupled with due diligence from said date to a subsequent reduction to practice or to the filing of the application. Original exhibits of drawings or records, or photocopies thereof, must accompany and form part of the affidavit or declaration of their absence satisfactorily explained. When a claim is rejected on grounds of a domestic patent which shows or describes but does not claim the invention, or on grounds of a foreign patent or printed publication which shows or describes the invention, the reference can be removed by affidavit provided the applicant can show facts of completion of the invention in this country before the reference date, and provided the reference date of the prior art is not more than one year before the date on which the application was filed in this country. The effective reference date of a United States patent is the filing date of the application on which the patent issued, the issue date of a foreign patent, and the publication date of a printed publication.

Prior art references most commonly removed by a Rule 131 affidavit involve rejections grounded on Section 102(a) of the Patent Code (printed publications, foreign and domestic patents), Section 102(e) (domestic patents), and Section 103 (obviousness).

In order to remove prior art references, the applicant must present facts which establish that he either reduced the invention to practice in this country before the reference date or he conceived the invention in this country and was diligent in reducing the invention to practice in this country from a date just before the reference date up to actual reduction to practice or to filing the application.

The second sentence of subsection (b) of Rule 131 sets forth the type of evidence required in the affidavit. The affidavit must state facts, and there must be documentary evidence in support of these facts. Documentary evidence may include drawings, research records, laboratory notebooks, etc., as discussed in Chapter 2. The Patent Office accepts presented evidence prima facie. However, if adversary proceedings such as interference or infringement litigation evolve, corroboration would be required.

Reduction to practice refers to the actual construction of the invention in a physical form. For a machine, this includes the actual building of the machine; for an article or composition, this includes the actual making of the article or composition; for a process, this includes the actual operation of the process.

Conception of the invention refers to the completion of the devising of the means for accomplishing the result. To demonstrate conception in the sense of Rule 131, one must have evidence such as drawings, records of a complete disclosure to another person, etc. In showing diligence in reduction to practice, one must have records of experimentation, records of buying material related to reduction to practice, etc.

There are certain references which cannot be removed as prior art under Rule 131. Any printed publication which has a publication date or any foreign patent which has an issue date more than one year prior to the applicant's effective filing date cannot be removed; this would constitute a statutory bar under Section 102(b). Any United States patent which has an effective filing date more than one year prior to the applicant's effective filing date cannot be removed; this would constitute a statutory bar under Section 102(e). Any reference that is the applicant's own foreign patent which claims the same invention and issues before the effective filing date of the application on a foreign application filed more than twelve months before the United States filing date cannot be removed; this is a statutory bar under Section 102(d). Any United States patent to another which substantially claims the same inventive concept cannot be removed; this involves interference.

Rule 132 provides for affidavits by which the applicant can traverse a prior art reference cited in a rejection. In essence, this rule provides a method by which the applicant can introduce relevant evidence to stress a point during prosecution. Rule 132 is cited below.

> When any claim of an application is rejected on reference to a domestic patent which substantially shows or describes but does not claim the invention, or on reference to a foreign patent or to a printed publication, or to facts within the personal knowledge of an employee of the office, or when rejected upon a mode or capability of operation attributed to a reference, or because the alleged invention is held to be inoperative or lacking in utility, or frivolous or injurious to public health or morals, affidavits or declarations traversing these references or objections may be received.

Rule 132 differs from Rule 131 in two respects. First, Rule 131 sets forth criteria for the facts presented in the affidavit: facts must be prima facie evidence as previously discussed; Rule 132 does not set forth such criteria. Second, Rule 131 sets forth rules determining when the presented facts overcome the rejection; Rule 132 does not do this, but only states that affidavits will be received.

Rule 132 affidavits are submitted most often to overcome obviousness rejections. The affidavit is often used to submit evidence of chemical utility.

In addition, the Rule 132 affidavit may be used to respond to a rejection on grounds of insufficient disclosure. The affidavit may contain evidence arguing that the original disclosure does indeed meet the Section 112 disclosure requirements.

ABANDONMENT

An application can become abandoned as a result of three circumstances: failure of the applicant to respond in timely fashion to an adverse office action, failure of the applicant to pay in timely fashion the issue fee, or a statement from the applicant expressing abandonment. This section is concerned on the one hand with the procedure for reviving abandoned applications once held abandoned by the Patent Office and on the other hand with the procedure for expressed abandonment as might be appropriate with continuing applications and defensive publications. Reviving applications for failure to pay the issue fee in timely fashion is discussed in a later chapter.

As previously discussed in this chapter, an application is held abandoned by the Patent Office if the applicant fails to respond within the time period set forth for response in an office action or if the applicant submits a response, filed in a timely fashion, which is not completely responsive. If the applicant wishes to revive the application after a notice of abandonment, he has two courses of action: he may submit a request for reconsideration of the ruling of abandonment or he may submit a petition for revival.

A request for reconsideration is appropriate only in cases where the application was held abandoned for the reason of improper response. In such cases, the applicant can submit a petition alleging that the response was indeed responsive, and if the examiner is so convinced, he might reverse his holding of abandonment.

The most common reason for abandonment is the failure to respond in timely fashion to an office action. The petition for revival of abandoned applications in this case is set forth in Patent Office Rule 137 cited below.

> An application abandoned for failure to prosecute may be revived as a pending application if it is shown to the satisfaction of the Commissioner that the delay was unavoidable. A petition to revive an abandoned application must be accompanied by a verified showing of the causes of the delay, by the proposed response unless it has been previously filed, and by the petition fee.

A verified showing consists of an affidavit from all persons involved in the circumstances resulting in abandonment. The decision to revive or not revive is based solely on the evidence in the affidavit as to whether the delay was unavoidable or not.

There are times when an applicant wishes to abandon a pending application. Such situations exist when filing a continuing application, when filing for a defensive publication, or when the applicant wishes to abandon the pending application and file a new application. Patent Office Rule 138 relates to the declaration of abandonment and is cited below.

> An application may be expressly abandoned by filing in the Patent and Trademark Office a written declaration of abandonment signed by the applicant himself and the assignee of record, if any, and identifying the application. Except as provided in sect. 1.262 an application may also be expressly abandoned by filing a written declaration of abandonment signed by the attorney or agent of record. Express abandonment of the application may not be recognized by the Office unless it is actually received by appropriate officials in time to act thereon before the date of issue.

The declaration of abandonment is a letter expressing abandonment signed by the applicant, his assignee, or the patent practitioner of record.

Chapter 8

PETITION AND APPEAL

INTRODUCTION

A rejection of a claim involves the substantive content of the claim; an objection to a claim involves the form of the claim. Rejections are based on statutory grounds, and as discussed earlier, a vast majority of all rejections are based on Sections 102 and 103 of the Patent Code. Objections are based on formal grounds, failure of some aspect of the application to follow Patent Office rules.

As a general rule, examiners' rejections can be appealed and examiners' objections can be petitioned. On the following pages are flow charts of the petition-appeal procedures. Petitions are directed to the commissioner of patents. An adverse decision from the commissioner may be reviewed in any federal district court as a civil action. An adverse decision from district court may be appealed to the appropriate circuit court of appeals and possibly to the Supreme Court if certiorari is granted.

A rejection by the primary examiner can be appealed to the Board of Appeals within the Patent Office. Adverse decisions from the Board of Appeals can be reviewed by either the Court of Customs and Patent Appeals or by the District Court of the District of Columbia, but not both. The Court of Customs and Patent Appeals is a constitutional court appealable to the Supreme Court. The appeal to the District Court for the District of Columbia is a civil action suit with appropriate appeals to the District of Columbia Circuit Court of Appeals and the Supreme Court.

APPEAL

The statutory provision for the first appeal is found in Section 134 of the Patent Code cited below.

> An applicant for a patent, any of whose claims has been twice rejected, may appeal from the decision of the primary examiner to the Board of Appeals, having once paid the fee for such appeal.

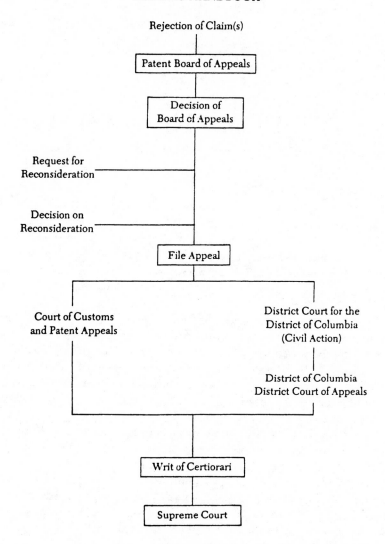

Flow chart 3. Appeal from adverse examiner's decision.

The first appeal of rejections is to the Patent Office Board of Appeals, and the appeal can be made after two rejections by the primary examiner regardless of whether the second rejection is final or nonfinal. The appeal is initiated by filing a notice of appeal which must be filed within the time set for response by the last office action. The notice of appeal must be accompanied by an appeal fee. After the notice of appeal, an appeal brief must be filed within two months or within the remaining time set for response by the last office action, whichever is longer. A second filing fee is required at the time of filing the brief.

Proceedings before the Board of Appeals can be conducted by the applicant if he is prosecuting his own application, or by his patent practitioner. The notice of appeal is best filed using a standard form provided by the Patent Office. This form serves to identify the application and to serve notice that the application is under appeal. It may be signed by either the applicant or his patent practitioner.

The brief is a more detailed response. By Patent Office Rule 142, the brief must contain the following information: "the authorities and arguments on which the appellant will rely to maintain his appeal, including a concise explanation of the invention which should refer to the drawings by reference, and a copy of the claims involved." The applicant must indicate at this time whether he wants an oral or nonoral hearing.

The appeal brief first goes to the examiner. The examiner studies the arguments and decides if he wants to allow the claims considering the appeal arguments or if he wants to proceed with the appeal. If he decides to appeal, the examiner prepares the "examiner's answer" in response to the appeal brief. A copy of the examiner's answer is supplied to the Board of Appeals and to the appellant.

Some art groups use the so-called "appeal conference" to help the primary examiner make the decision with regard to the appeal. This is an informal conference consisting of the primary examiner, an examiner charged with preparation of the examiner's answer, and another examiner knowledgeable in the art. These participants study the appeal brief, make their recommendation, and from this information the primary examiner makes his final decision. The primary examiner may decide to allow the claims, to negotiate the claims, or to allow appeal. It should be noted that not all art groups utilize the appeal conference, and within art groups that do use it, not all cases are handled by this method. This is at the discretion of the group director.

According to Section 1208 MPEP, the appellant should receive the examiner's answer within two months. If the examiner's answer contains new points of argument or new grounds of rejection, the appellant is entitled to file a "reply brief." The reply brief must be filed within twenty days from the date of the examiner's answer if it contains no new points of argument and within two months if it contains new grounds of rejection. In both cases, however, the reply brief must be directed only toward the new points of argument or toward the new rejection rather than toward the original brief in general. The reply brief is directed to the primary examiner.

After studying the reply brief, the examiner forwards the case to the Board of Appeals. Up to this point, all contact has been with the art group. When the file reaches the Board of Appeals, it is assigned an appeal number. If the applicant earlier asked for an oral hearing, it is placed in the appropriate docket, otherwise, it is placed in the nonoral docket. Cases are taken up by appeal number unless the case was made special at the beginning of prosecution or unless a petition is submitted and granted at this time to make the case special.

Board of Appeals

The composition of the Board of Appeals is dictated by Section 7 of the Patent Code cited below.

> The examiners-in-chief shall be persons of competent legal knowledge and scientific ability, who shall be appointed under the classified civil service. The Commissioner, the deputy commissioner, the assistant commissioners, and the examiners-in-chiefs shall constitute a Board of Appeals, which on written appeal of the applicant, shall review adverse decisions of examiners upon applications for patents. Each appeal shall be heard by at least three members of the Board of Appeals, the members hearing such appeal to be designated by the Commissioner. The Board of Appeals has sole power to grant rehearings. Whenever the Commissioner considers it necessary to maintain the work of the Board of Appeals current, he may designate any patent examiner of the primary examiner grade or higher, having the requisite ability, to serve as examiner-in-chief for periods not exceeding six months each.
>
> An examiner so designated shall be qualified to act as a member of the Board of Appeals. Not more than one such primary examiner shall be a member of the Board of Appeals hearing an appeal. The Secretary of Commerce is authorized to fix the per annum rate of basic compensation of each designated examiner-in-chief in the Patent and Trademark Office at not in excess of the maximum scheduled rate provided for positions in grade 16 of the General Schedule of the Classification Act of 1949, as amended. The per annum rate of basic compensation of each designated examiner-in-chief shall be adjusted, at the close of the period for which he was designated to act as examiner-in-chief, to the per annum rate of basic compensation which he would have been receiving at the close of such period if such designation had not been made.

The minimum number of members which can hear a case is three, and usually this is the number present. Typically the board is broken up in such a way that there are two or three boards sitting each day, one for chemical cases, one for mechanical cases, and one for electrical cases. By this method, cases can be processed at a faster pace, and those sitting have a knowledge in the field of art.

If the applicant requested an oral hearing in the notice of appeal, he or his representative is notified of the date and time for the hearing. The applicant is allowed two weeks to confirm his intent to appear; otherwise, the case is placed in the nonoral document. During the hearing the appellant is given about twenty minutes to state his case, his arguments are restricted in principle to the points of the brief or reply brief. The primary examiner is allowed about fifteen minutes if he requests to state his arguments.

If a nonoral hearing was originally requested, the briefs are assigned to a panel of the board for consideration. In these cases, the board makes its decision from the record.

After consideration, the board's decision is written by one member of the panel hearing the case; other members may concur or dissent, with or without opinion. Usually decisions are unanimous. The decision is usually handed down within two months and mailed to the appellant. The decision is not made public unless the appellant does so until a patent issues on the application.

The decision by the Board of Appeals may take four forms: (1) it may reverse the examiner in full or part, (2) it may sustain the examiner in full or part, (3) it may return the case to the primary examiner to consider a new ground proposed by the board for rejecting any allowed claim that the board believes should be considered, and (4) it may include a statement directing the allowance of a claim in an amended form, as amended to conform with this board's directive.

If the board sustains the rejection of any claim, the rejection may be based on grounds advanced by the examiner or it may be based on new grounds advanced by the board. If the rejection is based on new grounds, the appellant may request that the case be returned to the examiner for further prosecution, or he may request that the board reconsider the grounds for rejection. If the appellant chooses the latter, he must submit the request within thirty days from the original decision, and he may submit additional information to the board addressed to the new grounds for rejection. The Board of Appeals has sole power to grant rehearing.

If the appellant chooses to return the application to the examiner for further prosecution, he must file a timely response which is usually set at thirty days by the board. In this case, prosecution is open only to subject matter to which the new grounds of rejection were applied. The examiner is now bound by the new grounds and may allow the claims if the new rejection is overcome in his opinion.

As indicated earlier, the board may return the case to the primary examiner to consider "any grounds for rejecting any allowed claim that it believes should be considered." When this happens, the appellant is given a period of time, usually one month, to submit amendments or reasons to the primary examiner to avoid these grounds. The examiner may adopt or not adopt the new grounds for rejection suggested by the board. The examiner thereafter submits his decision to the board, and the board may maintain its original decision, or it may render a new decision with reference to these grounds.

The fourth possibility relating to the board's decision is that it may include explicit statements making claims allowable when the claims are amended to conform to the directive of the decision. This is set forth in Rule 196, which states that in the case of "an explicit statement that a claim may be allowed in amended form, applicant shall have the right to amend in conformity with such statement, which shall be binding on the primary examiner in the absence of new references or grounds of rejection." It should be stressed, however, that such amendments are not always that straight forward because even though the applicant is allowed to amend the claim to conform to the decision's statement, the examiner may then reject the amended claim on other grounds.

Appeal from Adverse Decision of Board of Appeals

From the Board of Appeals, one may appeal an adverse decision to either the Court of Customs and Patent Appeals or the District Court for the District of Columbia, but not both. The appeal to the Court of Customs and Patent Appeals is set forth in Section 141 of the Patent Code cited below.

> An applicant dissatisfied with the decision of the Board of Appeals may appeal to the United States Court of Customs and Patent Appeals, thereby waiving his right to proceed under section 145 of this title. A party to an interference dissatisfied with the decision of the board of patents interferences on the question of priority may appeal to the United States Court of Customs and Patent Appeals but such appeal shall be dismissed if any adverse party to such interference, within twenty days after the appellant has filed notice of appeal according to section 142 of this title, files notice with the Commissioner that he elects to have all further proceedings conducted as provided in section 146 of this title. Thereupon the appellant shall have thirty days thereafter within which to file a civil action under section 146, in default of which the decision appealed from shall govern the further proceedings in the case.

The Court of Customs and Patent Appeals is a constitutional court, and it differs from civil court in that it considers only the record as present in the Patent Office; it takes no additional testimony. It argues the case as present on the record in file. This court has advantages over a civil action in that it is a lot faster and less expensive. Appeal to the Supreme Court is possible if granted certiorari.

If one wishes to add additional testimony, he must appeal in a civil action in the District Court for the District of Columbia. This appeal is set forth in Section 145 of the Patent Code cited below.

> An applicant dissatisfied with the decision of the Board of Appeals may unless appeal has been taken to the United States Court of Customs and Patent Appeals, have remedy by civil action against the Commissioner in the United States District Court for the District of Columbia if commenced within such time after such decision, not less than sixty days, as the Commissioner appoints. The court may adjudge that such applicant is entitled to receive a patent for his invention, as specified in any of his claims involved in the decision of the Board of Appeals, as the facts in the case may appear and such adjudication shall authorize the Commissioner to issue such patent on compliance with the requirements of law. All the expenses of the proceedings shall be paid by the applicant.

The disadvantages to this appeal are that it is more expensive and takes longer. An adverse decision from district court may be appealed to the District of Columbia Court of Appeals. From here, one can appeal to the Supreme Court if granted certiorari. The Supreme Court does occasionally hear a patent case, most recently one involving genetic engineering of microorganisms (*Chakrabarty*).

PETITIONS

Petitions are directed to the commissioner of patents and are most often submitted to obtain review of an examiner's decision relating to Patent Office rules. The petition is also used to revive abandoned applications and to make applications special.

When seeking review of an examiner's objection, the petition is first reviewed by another official in the executive branch of the Patent Office. Often this is the examiner's group director from whom it is difficult to get a favorable ruling. Review of the decision on any petition may be further reviewed by filing another petition, unless the decision came from the commissioner himself. In essence, by further petitions one can advance adverse decisions up the executive ladder of the Patent Office to the deputy assistant commissioner. The commissioner reviews decisions of petitions only in very unusual circumstances.

If the decision is made to seek court review of an adverse decision from petition, such review can be made in any district court as a civil action. An adverse decision from district court can be appealed to the appropriate circuit court of appeals.

It should be noted that federal court review of a petition is made available under the Administrative Procedure Act. Under this act, there is a "presumption of administrative correctness" attached to the decision of the Patent Office. To obtain a favorable ruling, you must overcome this presumption of correctness. This is a difficult, almost impossible undertaking. Favorable rulings in federal court are rare.

Some of the more commonly filed petitions are listed below along with any required filing fee.

1. Petition against an examiner's decision to make a rejection final or the examiner's refusal to enter an amendment after final action under Rule 116.

2. Petition against examiner's decision on restriction requirement (except a Markush restriction, which is appealable).

3. Petition to obtain a filing date.

4. Petition to revive an abandoned application resulting from failure to respond in timely fashion to an office action.

5. Petition to revive an abandoned application for failure to pay issue fee.

6. Petition to make an application special so that the application is advanced out of its turn during examination.

Chapter 9

ALLOWANCE, ISSUE, AND CORRECTION

INTRODUCTION

At some point during the prosecution of the patent application, the claims are either allowed or they are finally rejected. The applicant is notified of this in an office action, and thereafter he is no longer entitled as a matter of right to make substantive amendments to his application. This action, in effect, serves to terminate prosecution of the application on its merits.

In the case of final rejection, there are three courses of action available to the applicant if he persists in further prosecuting the application: he may file an amendment in an effort to eliminate the grounds on which the rejection was based as discussed in Chapter 7; he may file a continuing application followed by abandonment of the rejected application as discussed in Chapter 4; or he may appeal the rejection as discussed in Chapter 8.

In the case of allowed claims, the applicant usually accepts the claims as allowed. However, if he persists in further prosecuting the application, he has two courses of action: he may file a continuing application followed by abandonment or issue of the allowed application or he may file an amendment under 37 CFR 1.312. Under this rule, the applicant is very limited in the scope of his amendment as discussed in this chapter.

This chapter is concerned with the application after allowance, amendments after notice of allowance and issue, and correction after issue.

ALLOWANCE

The first indication to the applicant that he has allowable claims is in the Patent Office action which indicates claims allowable and indicates that prosecution on merits is closed. This usually precedes the formal notice of allowance by two or three months. Substantive amendments to the application will not be entered after preliminary indication of allowance since prosecution on merits is closed. The Patent Office may enter amendments which do not expand the scope of the claims under Patent Office Rule 312, but this is not a matter of right but occurs at the discretion of the Patent Office. If the applicant wants to substantially change his application, he must file a continuing application followed by abandonment or issuance of the allowed application.

Following the office action by a couple of months is the formal notice of allowance. The notice of allowance is set forth in Section 151 of the Patent Code cited below.

> If it appears that applicant is entitled to a patent under the law, a written notice of allowance of the application shall be given or mailed to the applicant. The notice shall specify a sum, constituting the issue fee or a portion thereof, which shall be paid within three months thereafter.
>
> Upon payment of this sum the patent shall issue, but if payment is not timely made, the application shall be regarded as abandoned.
>
> Any remaining balance of the issue fee shall be paid within three months from the sending of a notice thereof and, if not paid, the patent shall lapse at the termination of this three-month period. In calculating the amount of a remaining balance, charges for a page or less may be disregarded.
>
> If any payment required by this section is not timely made, but is submitted with the fee for delayed payment and the delay in payment is shown to have been avoidable, it may be accepted by the Commissioner as though no abandonment or lapse had ever occurred.

The first paragraph of this statute calls for a formal notice of allowance to be sent to the applicant along with a sum constituting the issue fee. Between the office action and the mailing of the formal notice of allowance, the examiner subjects the application to a final review. The final review involves a final check for compliance with office rules, renumbering of allowed claims in sequence, and proofreading drawings, title, abstract, etc., for minor errors. In addition, the examiner does an interference search with other pending applications in related technology. If the same subject matter is being claimed in another application, the examiner may initiate interference proceedings as discussed in the following chapter.

After the notice of allowance has been mailed, the case is transferred from the examiner to the Patent Issue Division. Thereafter, the applicant has three months to submit amendments under Patent Office Rule 312 and to pay the issue fee to avoid abandonment.

Further prosecution of an application after notice of allowance is not a matter of right. Certain amendments may be entered, however, depending upon when the amendment is received and what the amendment proposes. Patent Office Rule 312 relates to such amendments and is cited below.

> Amendments after the notice of allowance of an application will not be permitted as a matter of right. However, such amendments may be made if filed not later than the date the issue fee is paid, on the recommendation of the

primary examiner, approved by the Commissioner, without withdrawing the case from issue.

The entry of amendments is at the discretion of the primary examiner. Usually the examiner will enter amendments only when it is demonstrated that the proposed amendment is necessary to provide adequate protection for the invention. As a general rule, amendments proposing to add claims broader than allowed claims will not be entered; amendments proposing claims which are narrower than allowed claims (dependent claims) will be entered if it is demonstrated that such claims are necessary to give adequate patent protection and that additional prior art searching will not be necessary.

ISSUE

The patent grant itself is a single-page, red-ribbon document issued to the patentee and his heirs or assigns in the name of the United States government under the seal of the Patent and Trademark Office. This document contains the title of the invention and the date of issue. Attached to and forming part of the patent are the specification and drawings. These are the working parts of the patent. On the date of issue, the record of the patent becomes open to the public. Printed copies of the specification and drawings also become available on that date, or shortly thereafter.

In cases where the application has been assigned (entire interest), the patent issues to the assignee of record. In cases where less than the entire interest has been assigned, the patent issues jointly to the inventor and assignee.

Patent rights begin on the date of issue and last for a seventeen-year term from this date. From the date of issue, the patentee has a two-year period to file a broader reissue application if he wishes as later discussed.

CORRECTION

For the most part when a patent issues, it is out of the jurisdiction of the Patent Office. One exception to this is that the Patent Office has methods for correcting certain mistakes, errors, or defects in issued patents. For these to be correctable, however, they must have occurred without deceptive intention.

Errors in issued patents may be minor in nature such as printing errors in the specification or drawings. These usually have little effect on the substance of the patent and are correctable by a certificate of correction. On the other hand, the patent may be defective in certain respects; it may claim too much, or it may claim too little. In these cases, the law provides that the patentee may apply for a reissue patent. The corrective procedure that one uses for correcting a defective patent depends on the type of error involved and the magnitude of the error.

There are three statutory methods for correcting issued patents: certificates of correction used to correct minor errors or omissions and to correct inventorship; reissue applications used to correct substantive errors in the patent; and disclaimers used to eliminate claims from the patent or to shorten the life of the patent.

CERTIFICATE OF CORRECTION

The certificate of correction is used to correct minor mistakes in issued patents such as misspelled words, typographical errors, incorrect formulas, etc. These are mistakes that can be corrected without changing the substantive content of the patent. Changes by the certificate of correction may not incorporate new matter into the patent, nor may changes require a substantive reexamination of the patent by the Patent Office.

The certificate of correction itself is a separately typed page which is attached to the issued specification setting forth the error or errors wherever they occur. The correction has the same effect as if the original patent had been issued in the corrected form.

If the error in the issued patent was the fault of the Patent Office, correction by the certificate of correction might be obtained without charge as set forth in Section 254 of the Patent Code cited below.

> Whenever a mistake in a patent, incurred through the fault of the Patent and Trademark Office, is clearly disclosed by the records of the Office, the Commissioner may issue a certificate of correction stating the fact and nature of such mistake, under seal, without charge, to be recorded in the records of patents. A printed copy thereof shall be attached to each printed copy of the patent, and such certificate shall be considered as part of the original patent. Every such patent, together with such certificate, shall have the same effect and operation in law on the trial of actions for causes thereafter arising as if the same had been originally issued in such corrected form. The Commissioner may issue a corrected patent without charge in lieu of and with like effect as a certificate of correction.

This statute relates to errors that occur between the application as prosecuted and the printed form of the issued patent. Most errors of this nature occur during the printing process, and these errors may include omitted letters, transposed letters, omitted words, incorrect printing of a formula, and other related printing errors. Such printing errors may have no effect on the text of the patent. On the other hand, they may be located at a point such as a claim, where they affect the legal interpretation of the patent. In either case, the patentee may apply for a certificate of correction.

The certificate of correction is not a matter of right but is provided at the discretion of the commissioner: "the Commissioner may issue a certification of correction." To initiate the certificate of correction, the patentee simply notifies the Patent Office in writing of the error. Whether the office issues the certificate depends on the magnitude of the error. If the error has any effect on the readability of the patent, the certificate is usually issued. If the error is so minor it has no effect, the office usually makes the letter of record but does not issue the certificate.

If the error is not the fault of the Patent Office but that of the applicant, correction by the certificate of correction can be obtained by paying a required fee. This provision is set forth in Section 255 of the Patent Code cited below.

> Whenever a mistake of a clerical or typographical nature, or of minor character, which was not the fault of the Patent and Trademark Office, appears in a patent and a showing has been made that such mistake occurred in good faith, the commissioner may, upon payment of the required fee, issue a certificate of correction, if the correction does not involve such changes in the patent as would constitute new matter or would require reexamination. Such patent, together with the certificate, shall have the same effect and operation in law on the trial of actions for causes thereafter arising as if the same had been originally issued in such corrected form.

The type of errors correctable by this statute are misspelled words, incorrect formulas, typographical errors, etc., originating with the patentee. These are errors that occur in the application as well as the issued patent. In addition, the certificate of correction is used to correct inventorship as discussed in a later section of this chapter.

To obtain a certificate of correction, the patentee must submit a request including a statement that the error occurred without deceptive intent, a statement that the requested correction does not involve new matter and no reexamination would be required, and a fee. This information is forwarded to the Patent Issue Division, which has responsibility for determining if an error has been made and whether the error is of such nature to justify correction. The *Official Gazette* contains a weekly list of patents for which certificates of correction have been issued.

REISSUE PATENTS

When an issued patent is defective in certain respects, the patent statutes provide that the patentee may apply for a reissue patent. In applying for reissue, the defective patent is in effect surrendered to the Patent Office, a reissue application is submitted and completely reexamined, and a reissue patent is granted to replace the defective one for the balance of the unexpired term of the original patent. The statutory provision for reissue is set forth in Section 251 cited below.

> Whenever any patent is, through error without any deceptive intention, deemed wholly or partly inoperative or invalid, by reason of a defective specification or

drawing, or by reason of the patentee claiming more or less than he had a right to claim in the patent, the Commissioner shall, on the surrender of such patent and the payment of the fee required by law, reissue the patent for the invention disclosed in the original patent, and in accordance with a new and amended application, for the unexpired part of the term of the original patent. No new matter shall be introduced into the application for reissue.

The Commissioner may issue several reissued patents for distinct and separate parts of the thing patented, upon demand of the applicant, and upon payment of the required fee for a reissue for each of such reissued patents.

The provisions of this title relating to applications for reissue of a patent, except that application for reissue may be made and sworn to by the assignee of the entire interest if the application does not seek to enlarge the scope of the claims of the original patent.

No reissued patent shall be granted enlarging the scope of the claims of the original patent unless applied for within two years from the grant of the original patent.

Defects may be corrected by reissue if the uncorrected defect would make the patent partly or wholly inoperative or invalid. The defect most commonly corrected by reissue involves cases where the patentee claimed more than he had a right to claim or cases where the patentee claimed less than he had a right to claim. In each case, error must have occurred without deceptive intention.

In contrast to the certificate of correction, which is at the commissioner's discretion, reissue in this case is a matter of right: "the Commissioner shall, on the surrender of such patent and the payment of fee required by law, reissue the patent for the invention disclosed in the original patent." Reissue proceedings are initiated by a petition from the patentee stating the reasons why he believes his patent to be inoperative or invalid. With the petition and application for reissue, there must be an offer to surrender the original patent. However, the Patent Office usually does not request the patent unless and until the reissue patent is granted. At this time, the ribbon copy of the patent must be surrendered.

When applying for reissue, the application for reissue must comply with the disclosure in the original patent. If the reissue application seeks to enlarge the scope of the original claims, it must be filed within two years after issuance of the original patent. If the reissue application seeks to narrow the scope of the original claims, it may be filed anytime during the life of the parent patent so long as it is done in a reasonable time after the defect is discovered. The most common narrowing reissue occurs when the patentee discovers additional prior art which encompasses a claim or claims.

Reexamination

The reexamination by the Patent Office of the reissue application involves a complete substantial reexamination of all claims, both amended claims and claims carried over unchanged from the original patent. All claims in the reissue application may be rejected on either prior art that was previously considered or on newly discovered prior art. Therefore, the reissue application lays open the entire property right for reexamination. With the reissue application, the patentee may end up with nothing, or he may end up with a reissue patent which enhances the scope of his patent protection of the invention. With a broadened reissue patent, the presumption of validity is strengthened because the reissue has been twice examined. On the other hand, failure to obtain a narrowed reissue makes the existing patent suspect to invalidation and thus infringement.

In contrast to the original application, the files of the reissue application are open to the public. A listing of reissue applications is printed weekly in the *Official Gazette*. Interested parties may submit prior art or any other information to the Patent Office to be considered during the reexamination procedure.

The time involved in examining a reissue application is usually less than for an original application because the reissue is made special by Patent Office Rule 176 as previously discussed in Chapter 6. The reissue application is therefore advanced out of the normal order of examination and processed at an accelerated rate.

Certain defects cannot be corrected by reissue. The most prominent of these occurs when the original application does not contain an enabling disclosure as required by Section 112 of the Patent Code. In this case, the only alternative is to file a new application in which the new disclosure is entitled to the filing date of the new application.

Effect on Infringement

Section 252 of the Patent Code relates to the effect of reissue on the original patent, the effect of reissue on infringement of claims common to both the original and reissue patent and to the effect of reissue on infringement of claims not common to the original and reissue patent, and is cited below.

> The surrender of the original patent shall take effect upon the issue of the reissued patent, and every reissued patent shall have the same effect and operation in law, on the trial of actions for causes thereafter arising, as if the same had been originally granted in such amended form, but in so far as the claims or the original and reissued patents are identical, such surrender shall not affect any action then pending nor abate any cause of action then existing, and the reissued patent, to the extent that its claims are identical with the original patent, shall constitute a continuation thereof and have effect continuously from the date of the original patent.
>
> No reissued patent shall abridge or affect the right of any person or his successors in business who made, purchased or used prior to the grant of a reissue anything patented by the reissued patent, to continue the use of, or to sell to

others to be used or sold, the specific thing so made, purchased or used, unless the making, using or selling of such thing infringes a valid claim of the reissued patent which was in the original patent. The court before which such matter is in question may provide for the continued manufacture, use or sale of the thing made, purchased or used as specified, or for the manufacture, use or sale of which substantial preparation was made before the grant of the reissue, and it may also provide for the continued practice of any process patented by the reissue, practiced, or for the practice of which substantial preparation was made, prior to the grant of the reissue, to the extent and under such terms as the court deems equitable for the protection of investments made or business commenced before the grant of the reissue.

When the reissue patent issues, the original patent is surrendered. The reissue patent has the same effect as if it had been originally granted in the amended form with regard to claims common to both the original and reissue patents. For these common claims, reissue changes nothing as far as infringement is concerned. The reissue is in effect a continuation of the original patent.

When the reissue patent enlarges the scope of the claims of the original patent, however, intervening rights may come into play. Intervening rights are provided for in the second sentence of Section 252. This sentence relates to activity which infringes the broadened reissue claims, but does not infringe any claims in the original patent. In effect, this sentence provides for the resale or continued use of things which infringe claims of the reissue patent but not the claims of the original patent. These rights allow one to use and sell royalty-free anything protected by the reissue patent which was made or acquired before the reissue was granted.

The third sentence of Section 252 relates to intervening rights of continued manufacture. In certain cases where the manufacturer has spent large amounts of capital for facilities, equipment, etc., to set up to manufacture a product or use a process covered by a broadened reissue claim (but not covered by claims in the original patent), the court may provide for the continued manufacture of the product or use of the process. The intervening right of continued manufacture is not a matter of right by this statute, but the court may provide for continued manufacture in certain cases.

DISCLAIMER

The third method of correcting errors in issued patents is by a disclaimer. A "substantive disclaimer" can be used to remove invalid or defective claims from a patent, and this may be appropriate when one discovers a prior art reference subsequent to examination of the application which anticipates certain claims in the patent. A "terminal disclaimer" can be used to shorten the term of a patent, and this type disclaimer would be appropriate when one desires to dedicate the terminal period of his patent to the public. In addition, the terminal disclaimer might be used to overcome a double patenting rejection.

The disclaimer is set forth in Section 253 of the Patent Code cited below.

> Whenever, without any deceptive intention, a claim of a patent is invalid the remaining claims shall not thereby be rendered invalid. A patentee, whether of the whole or any sectional interest therein, may, on payment of the fee required by law, make disclaimer of any complete claim, stating therein the extent of his interest in such patent. Such disclaimer shall be in writing and recorded in the Patent and Trademark Office, and it shall thereafter be considered as part of the original patent to the extent of the interest possessed by the disclaimer and by those claiming under him.
>
> In like manner any patentee or applicant may disclaim or dedicate to the public the entire term, or any terminal part of the term, of the patent granted or to be granted.

The first sentence of the statute relates to the substantive disclaimer. This type of disclaimer may be used to delete selected claim(s) from the patent. This most often involves deleting claims which are anticipated or made obvious by a prior art reference found subsequent to examination of the application. As with other types of correction, the error must have occurred without deceptive intention. Invalidity is not a prerequisite for deleting claims by the procedure, however. For example, the patentee may dedicate certain claims to the public by this procedure while retaining rights to other claims.

Patent Office Rule 321 relates to the disclaimer and is cited below.

> (a) A disclaimer under 35 U.S.C. 253 must identify the patent and the claim or claims which are disclaimed, and be signed by the person making the disclaimer, who shall state therein the extent of his interest in the patent. A disclaimer which is not a disclaimer of a complete claim or claims may be refused recordation. A notice of the disclaimer is published in the Official Gazette and attached to the printed copies of the specification. In like manner any patentee or applicant may disclaim or dedicate to the public the entire term, or any terminal part of the term, of the patent granted or to be granted.
>
> (b) A terminal disclaimer, when filed in an application to obviate a double patenting rejection, must include a provision that any patent granted on that application shall be enforceable only for and during such period that said patent is commonly owned with the application or patent which formed the basis for the rejection.

A terminal disclaimer relates to all the claims in a patent in contrast to the substantive disclaimer, which relates to select claims. The terminal disclaimer may be used to dedicate all claims in a patent to the public. It is sometimes used by nonprofit research organizations, the government, or companies trying to avoid an antitrust situation. The terminal disclaimer can be worded to become effective immediately or at some given time in the future.

The terminal disclaimer is also used to overcome a double patenting rejection. When the examiner asserts that an application is claiming the same inventive concept as claimed in a previous patent to the same invention, the rejection may be overcome by a terminal disclaimer of the application. Any patent issuing from the application is enforceable only for the time of the previously issued, commonly owned patent.

CORRECTION OF INVENTORSHIP

Application for patent must be made by the inventor and determination of inventorship was discussed in Chapter 2. Any patent identifying one as an inventor who is not an inventor or any patent not including all co-inventors is invalid unless corrected. There are statutory procedures for correcting inventorship in both the application and the issued patent provided the error is a nonjoinder or a misjoinder and the error occurred without deceptive intention. Title 35, United States Code, Section 116, relates to correction of inventorship involving applications, and is cited below in pertinent part.

> Whenever a person is joined in an application for patent as a joint inventor through error, or a joint inventor is not included in an application through error and such error arose without deceptive intention on his part, the Commissioner may permit the application to be amended accordingly, under such terms as he prescribes.

Section 256 relates to correction of inventorship for issued patents and is cited below.

> Whenever a patent is issued on the application of persons as joint inventors and it appears that one of such persons was not in fact a joint inventor, and that he was included as a joint inventor by error and without any deceptive intention, the Commissioner may, on application of all the parties and assignees, with proof of the facts and such other requirements as may be imposed, issue a certificate deleting the name of the erroneously joined person from the patent.
>
> Whenever a patent is issued and it appears that a person was a joint inventor, but was omitted by error and without deceptive intention on his part, the Commissioner may, on application of all the parties and assignees, with proof of the facts and such other requirements as may be imposed, issue a certificate adding his name to the patent as a joint inventor.
>
> The misjoinder or nonjoinder of joint inventors shall not invalidate a patent, if such error can be corrected as provided in this section. The court before which such matter is called in question may order correction of the patent on notice and hearing of all parties concerned and the Commissioner shall issue a certificate accordingly.

Both statutes provide for correction in cases of nonjoinder, where an inventor(s) was erroneously omitted, and for cases of misjoinder, where an inventor(s) was erroneously included. Neither statute provides for correction in cases where none of the true inventors are designated. For correction to be available by either statute, the omission must have occurred by error without deceptive intention. An application is corrected by amendment, and a patent is corrected by a certificate of correction.

Chapter 10

INTERFERENCE

INTRODUCTION

Occasionally two or more applications or one or more applications and one or more patents filed by different inventors claim substantially the same patentable invention. Since a valid patent can be granted to only one inventor, a proceeding known as "interference" is instituted to determine the first inventor and entitlement to the patent. The parties involved in the proceeding submit evidence of facts showing when they made the invention, and the question of priority is determined by the Board of Patent Interference. From the decision of the Board of Patent Interference, the losing party may appeal to the Court of Customs and Patent Appeal or file a civil action in the appropriate district court. About one percent of all applications become involved in interference.

The subject area of interference is complex; it involves affidavits, motions, testimony, and other legal maneuvers. The objective of this chapter is to present an introduction to the early stages of application-application and application-patent interference, the stages in which the inventor is likely to be involved. Patent-patent interference is defined by 35 U.S.C. Sec. 291, which is not discussed in this text.

STATUTORY PROVISION

The statutory provision for interference is set forth in Section 135 of the Patent Code cited below.

> (a) Whenever an application is made for a patent which, in the opinion of the commissioner, would interfere with any pending application, or with any unexpired patent, he shall give notice thereof to the applicants, or applicant and patentee, as the case may be. The question of priority of invention shall be determined by a board of patent interferences (consisting of three examiners of interferences) whose decision, if adverse to the claim of an applicant, shall constitute

the final refusal by the Patent and Trademark Office of the claims involved, and the Commissioner may issue a patent to the applicant who is adjudged the prior inventor. A final judgment adverse to a patentee from which no appeal or other review has been or can be taken or had the patent, and notice thereof shall be endorsed on copies of the patent thereafter distributed by the Patent and Trademark Office.

(b) A claim which is the same as, or for the same or substantially the same subject manner as, a claim of an issued patent may not be made in any application unless such a claim is made prior to one year from the date on which the patent was granted.

(c) Any agreement or understanding between parties to an interference, including any collateral agreements referred to therein, made in connection with or in contemplation of the termination of the interference, shall be in writing and a true copy thereof filed in the Patent and Trademark Office before the termination of the interference as between the said parties to the agreement or understanding. If any party filing the same so requests, the copy shall be kept separate from the file of the interference, and made available only to Government agencies on written request, or to any person on a showing of good cause. Failure to file the copy of such agreement or understanding shall render permanently unenforceable such agreement of understanding and any patent of such parties involved in interference or any patent subsequently issued on any application of such parties so involved. The Commissioner, may, however, on a showing of good cause for failure to file within the time prescribed, permit the filing of the agreement or understanding during the six-month period subsequent to the termination of the interference as between the parties to the agreement or understanding.

The Commissioner shall give notice to the parties or their attorneys of record, a reasonable time period to said termination, of the filing requirement of this section. If the Commissioner gives such notice at a later time, irrespective of the right to file such agreement or understanding within the six-month period on a showing of good cause, the parties may file such agreement or understanding within sixty days of the receipt of such notice.

Any discretionary action of the Commissioner under this subsection shall be reviewable under section 10 of the Administrative Procedure Act.

When it is the opinion of the Patent Office that a submitted application would interfere with another pending application or with an unexpired patent, the office may initiate interference to determine priority. Interference may be initiated when the application claims the same invention, or substantially the same invention, as another pending application or unexpired patent.

Interference may be initiated by the primary examiner at his own discretion or by the applicant's suggestion. Application-application interference is usually initiated by the examiner; patent-application interference is usually initiated by the applicant. This most commonly occurs when the applicant has an application pending and a patent issues to another claiming the same subject matter. In this case, the applicant may initiate proceedings as discussed later in this chapter. However, interference will be granted only if the patent has not been issued for more than one year prior to the filing of the conflicting application and when the application has patentable subject matter.

The claims in question during interference are called "counts." Usually the count corresponds exactly to a claim contained in the application or patent of one of the parties involved. However, when the contesting parties do not have identical phraseology in their claims directed to an identical inventive concept, the examiner may suggest counts representing the interfering subject matter.

Because of the expense of an interference proceeding, settlement is often reached by the parties involved prior to settlement by the Board of Interference. In these cases, the parties exchange evidence, determine priority on their own, and the winner licenses the losing party of the resulting patent. According to the above statute, the settlement agreement must be filed in the Patent Office. Failure to file such agreement makes the agreement unenforceable. The reasoning behind this is to prevent the Patent Office from continuing a case which has already been settled.

INTERFERENCE LAW

Interference law is substantially based on subsection (g) of Section 102 of the Patent Code cited below.

> (g) Before the applicant's invention thereof the invention was made in this country by another who had not abandoned, suppressed, or concealed it. In determining priority of invention there shall be considered not only the respective dates of conception and reduction to practice of the invention, but also the reasonable diligence of one who was first to conceive and last to reduce to practice, from a time prior to conception by the other.

The party with priority is the first party to reduce the invention to practice, unless a second party first conceived the invention and was continuously diligent in his efforts toward reduction to practice of the invention from a time just before the first party's conception of the invention until the second party actually reduced the invention to practice, or unless the first party to reduce the invention to practice either abandoned, suppressed, or concealed the invention. Interference law is based substantially on the above sentence.

Incorporated into interference law are the concepts of conception, reduction to practice, diligence, abandonment, suppression, and concealment, most of which were discussed in this context in Chapter 2. In summary, conception is the process of forming the ideal in the mind to the point that the invention would be workable if reduced to practice. Reduction to practice refers to the construction of the invention in a physical form which is workable. In the case of machines, this includes the actual building of the machine; in the case of chemical compositions, this includes the actual making of the composition; and in the case of a process, this includes setting up the process to the point of production.

Diligence involves an effort toward reducing the ideal to a physical form which is workable. This effort might involve experimentation, or it might involve collecting components, etc. The important thing is that an effort is being made to actually reduce the invention to practice. This effort is interpreted as evidence that one is working toward the filing of a patent application.

Abandonment has the same meaning as discussed in Chapter 2. Suppression or concealment involves keeping the invention secret after reduction to practice; the inventor does not make a disclosure to the public. In this case, it is assumed that one is not working toward filing a patent application and potential patent rights are lost.

INITIATION

As previously indicated, interference can be initiated either by the examiner suggesting claims to one or both parties to copy, or by one party copying claims from another party's application or patent on his own initiative. The examiner usually initiates application-application interference, although he may initiate patent-application interference if he finds an interfering patent having a filing date later than that of an application. The applicant usually initiates patent-application interference, most often when a patent issues to another claiming the same invention as his pending application.

Suggested Claims

When an examiner observes that two or more applications, one of which is in condition for allowance, are claiming the same invention, he may set forth one or more claims to be copied by one party for the purpose of interference. When initiating interference between applications, the examiner tries to find one or more claims present in one of the applications to suggest to the other party to copy. When it is not possible to select suitable counts, that is, claims with similar phraseology directed to an identical inventive concept, the examiner will frame a claim or claims which read on the applications involved and clearly express the interfering subject matter.

When an applicant to whom claims have been suggested refuses to copy the claim for the purpose of interference, it is deemed by the examiner that the subject matter has been disclaimed by the applicant. In these cases, the claims in the application are rejected as being anticipated by the other application.

Usually an examiner will not suggest claims for interference if the pending applications differ more than three months in their effective filing date. The reason for this policy is that the junior party so seldom wins interference, it is unfair to the senior party to delay his patent rights for the time involved in the proceeding. In cases which differ by more than three months, the claims of the senior party are allowed and the claims of the junior party are rejected on the basis of the senior party's patent. At this point, the junior party can initiate interference by copying the senior party's allowed claims by the following procedure.

Voluntarily Copying Claims

If while an applicant has an application pending, he observes a claim in a recently issued patent, or for that matter knows of a claim in another application, he may copy the claim from the application or patent to initiate interference. However, when copying claims from a patent, the interference must be initiated within one year of the issuance of the patent. This rule is set forth in Section 135, previously cited.

When an applicant copies claims from an application, he must identify the other application in accordance with Patent Office Rule 203(d); when an applicant copies claims from a patent, he must identify the patent in accordance with Patent Office Rule 205(b). This is to prevent the Patent Office from issuing patents with identical claims.

The applicant is not granted interference just because he copies claims; he must submit support for the copied claims. If the applicant is the junior party, that is, if he has a filing date after the other party, he must submit affidavits with evidence of an invention date prior to the other party's filing date.

According to Rule 204, if the applicant is less than three months junior to the patentee, he only has to file a statement that he made his invention in this country before the patentee's effective filing date. If the applicant is more than three months junior, the affidavit must be more elaborate; the affidavit must set forth documentary evidence relating to the facts, and when possible, these facts should be corroborated.

When the examiner receives an indication that an applicant has copied claims, he reviews the affidavits for form, but not for substance. As discussed hereafter, it is the Board of Patent Interference that declares interference.

DECLARATION

While it is the primary examiner who determines the existence of interference, either by his initiative or the applicant's initiative, it is a patent interference examiner who declares the interference. The primary examiner forwards the case file to the Board of Interference, indicating which counts he believes should be contested in the interference. If in the opinion of the interference examiner sufficient basis for interference exists, he then declares interference.

When interference is declared, the next actions by the applicant are those of filing a preliminary statement and filing motions. The preliminary statement must be filed within two months, and motions must be filed within four months after declaration of interference.

The Preliminary Statement

The preliminary statement is a statement by the inventor setting forth facts about the invention. A copy of the preliminary statement is served to each opposing party, provided they also file a preliminary statement as later discussed. The contents of the statement are detailed in Rule 216 and Rule 217. Rule 216, which concerns inventions made in the United States is cited below.

> (a) The preliminary statement must state that the party made the invention set forth by each count of the interference, and whether the invention was made in the United States or abroad. When the invention was made in the United States the preliminary statement must set forth as to the invention defined by each count the following facts:
>
> (1) The date upon which the first drawing of the invention was made; if a drawing of the invention has not been made prior to the filing date of the application, it must be so stated.
>
> (2) The date upon which the first written description of the invention was made; if a written description of the invention has not been made prior to the filing date of the application, it must be so stated.
>
> (3) The date upon which the invention was first disclosed to another person; if the invention was not disclosed to another person prior to the filing date of the application, it must be so stated.
>
> (4) The date of the first act or acts susceptible of proof (other than making a drawing or written description or disclosing the invention to another person) which, if proven, would establish conception of the invention, and a brief description of such act or acts; if there have been no such acts, it must be so stated.
>
> (5) The date of the actual reduction to practice of the invention; if the invention has not been actually reduced to practice before the filing date of the application, it must be so stated.
>
> (6) The date after conception of the invention when active exercise of reasonable diligence toward reducing the invention to practice began.
>
> (b) When an allegation as to the first drawing (paragraph (a)(1) of this section) and/or as to the first written description (paragraph (a)(2) of this section) is made, a copy of such drawing and/or written description must be attached to the statement. (See sect. 1.223(c).)
>
> (c) If a party intends to rely solely on a prior application, domestic or foreign, and on no other evidence, the preliminary statement may so state and need not be signed or sworn to or declaration made by the inventor.

Patent Office rule 217 relates to inventions made abroad and is cited below.

> When the invention was made abroad the facts specified by sect. 1.216(a) (1) to (6) are not required, and in lieu thereof there should be stated:
> (1) When the invention was introduced into this country by or on behalf of the party, giving the circumstances with the dates connected therewith which are relied upon to establish the fact and, when appropriate, including allegations of activity in this country of the nature of that represented by sect. 1.216(a) (1) to (6) and documentary attachments if the allegations relate to a drawing or written description. Such statement may be signed and sworn to, or made in the form of a declaration, either by the inventor or by one authorized to make the statement and having knowledge of the facts alleged therein.
> (2) If a party is entitled to the benefit of the second sentence of 35 U.S.C. 104, he must so state and his preliminary statement must include allegations of activity abroad corresponding to those required by sect. 1.216 (a) (1) to (6).

If a party fails to file a preliminary statement, his date of invention is restricted to his effective filing date. When a party does file a preliminary statement, he must notify all opposing parties of that fact and he must send a copy to all opposing parties from whom a notification of filing a statement has been received. If a party does not file a preliminary statement, he is not entitled to receive copies of the preliminary statement of other parties. The exception to this is the senior party, who can obtain copies of other preliminary statements when the senior party does not file a preliminary statement.

Motions

The motion period begins at the declaration of interference and lasts for four months. Any party to the interference may file motions seeking to do the following as summarized from Rule 231:

(1) To dissolve one or more counts of the interference
(2) To add or substitute new counts to the interference
(3) To substitute another application in the proceeding owned by the applicant
(4) To be accorded the benefit of an earlier filed application
(5) To amend an involved application by correcting inventorship

Each party is entitled to file an opposition to the motion, and these must be filed within twenty days after the motion period. These motions are decided by the primary examiner.

DETERMINING PRIORITY

Eventually the Board of Patent Interference determines priority of the invention from the evidence submitted. It should be noted that the board does not consider the question of patentability, only the question of priority; patentability is determined by the primary examiner.

The burden of proof rests with the junior party. The parties are presumed to have made their inventions in the chronological order of their filing dates; therefore, the burden of proof rests on the party seeking to establish an earlier date, the junior party.

Following the motion period is the "testimony period." This is the period in which evidence is submitted concerning the case. Evidence is usually in the form of sworn oral depositions and affidavits. The junior party first submits its evidence, and this is followed by the senior party submitting its evidence or rebuttal. After the senior party, the junior party gets another opportunity for rebuttal.

Following the testimony period, the parties submit briefs. After reviewing the briefs, the Board of Patent Interference holds the "final hearing." The board consists of three members, and if requested, they hear oral arguments. After considering the evidence, they make a decision in the form of a written opinion delivered by one member; others can concur, with or without opinion.

From the decision of the Board of Patent Interference, the losing party may appeal to the Court of Customs and Patent Appeal, or he may file a civil action in the appropriate district court.

Chapter 11

PATENTS AS INFORMATION TOOLS

INTRODUCTION

Patents are a major information source for new products, processes, and uses in every area of technology. Over one million patent documents are published each year worldwide. This includes issued patents, published unexamined applications, inventor's certificates, utility models, and utility certificates issued by some countries.

When an inventor submits an application for patent rights, he must describe his invention in the application. Patent documents therefore contain an immense amount of prior art information. With a little ingenuity and knowledge in the field, one can usually comprehend the invention and duplicate the examples disclosed in the patent document. This is particularly true of United States patents because the United States patent must by statutory mandate contain an "enabling disclosure" of the invention: the patent document must describe the invention, it must teach how to make and use the invention, and it must set forth the best method for making the invention.

Examples in foreign patent documents can be more difficult to comprehend and duplicate because the disclosure requirements of foreign patent offices are not as demanding as those of the United States Patent Office. If one is knowledgeable in the field, however, a lot of information can be obtained from foreign patent documents.

Many believe that patent literature has in recent years replaced conventional scientific and trade journals as the most up-to-date source of information on technological progress in a number of areas of technology for the following reasons:

(1) Industrial concerns usually do not publish information in journals or books until they have full patent coverage.

(2) A lot of information disclosed in patents is never published in other forms.

(3) Often patent documents describing the same inventions are published by several countries and are therefore available in different languages.

(4) Many countries publish their patent applications before examination; these documents are therefore published within a few months after filing.

While patent documents are available as an information source, they are not widely used for this purpose. This is the case because many consider the patent to be only a legal document, not an information source, and most people are not familiar with the classification systems which separate and store documents according to subject matter. The classification systems make documents in any technical area easier to locate and retrieve.

This chapter is concerned with the general aspects of the patent document as an information tool. Following chapters discuss how to find and distinguish information in a specific area in both domestic and foreign patent collections.

UNIQUE FEATURES OF PATENT DOCUMENTS

Since information searching is less expensive than experimenting, one should utilize all information sources available, including patent documents, technical articles, and trade journals. As a general rule, patent documents concentrate on the subject matter of the invention; technical and trade journal articles tend to be more general. They cover general subject matter rather than a single inventive entity. The patent collection as a prior art source therefore offers certain advantages not found in technical journals.

Searching the patent collection permits one to evaluate an ideal quickly and accurately. Patent documents are classified by the Patent Office into classes and subclasses which can be retrieved. By studying the classification files, one can do a "state of the art" study quickly and accurately. In addition, the Patent Office has digests of subject matter in many subject areas.

Searching the patent collection tells one whether an ideal belongs to another. The patent claims set forth the property rights of the patentee. By evaluating the claims, one can determine if a new product ideal would infringe another's property rights.

The patent collection is a good source for ideals. Even if an ideal is old and patented, becoming aware of it may be useful. An old invention might be given a new use, or it might be improved for a new product. Also, one might become aware of a component which would make a system work. Since patent rights can be licensed or sold, the component could be obtained to complete the system.

The disclosure in issuing patents indicates where competitors are concentrating their research activities. By knowing this, companies can evaluate their own competitive position in the market.

Patent documents often contain information which does not form part of the invention, but can be useful to the searcher. For example, patent documents reference related prior art, including domestic and foreign patents as well as technical publications.

PUBLICATION OF PATENT DOCUMENTS

The time at which patent documents are published varies in different countries. Some countries publish their patent applications before examination; as a result, applications in these countries are published within a few months after filing. This policy has the advantage of disclosing the technology in an application at a quicker rate. These countries, which are referred to as "quick issue" countries, are listed below.

Belgium	Netherlands
Brazil	Norway
Denmark	Portugal
Finland	South Africa
France	Sweden
Germany	United Kingdom
Japan	

Other countries publish applications after examination. Since the examination process takes time, these documents are published at a slower rate. These countries, which are referred to as "slow publishing" countries, are listed below.

Austria	Italy
Canada	Rumania
Czech Republic	Russia
Hungary	Switzerland
Israel	United States

In the United States, applications are kept secret during the examination process. The average time for the examination process is around 18 months, but, this varies from class to class. Some applications issue quicker, others take longer. When an application does not issue to a patent, it usually is not publicly disclosed.

Of the quick issue countries, Germany and Japan are probably the most prolific sources of prior art information. Unexamined German patent applications are published weekly, usually within six months of filing. Since many worldwide industrial concerns seek patent rights in Germany, a lot of technology is on file in Germany. Often equivalent applications for the same invention are filed in other countries. By following German applications, one can obtain advanced knowledge of this technology. Examined German applications are published about ten years after filing.

Japan is also very active in research, particularly chemical and electronic research. A lot of information is disclosed by Japanese patent documents, and it is disclosed quickly because Japan publishes patent applications before examination. While there are at times translation problems, these documents are available.

TYPE OF SEARCHES

There are several circumstances under which a patent search would be needed, including the following:

1. When making a decision whether to file an application (patentability search)
2. When one wishes to be apprised of all prior art in a technical field or to learn of an existing solution to a problem (state of the art search)
3. Prior to initiating an infringement procedure or during infringement as a defendant (infringement search)
4. When assessing the value of a competitor's patent which is preventing free marketing of a product or process (validity search)
5. To assist in managerial decisions such as determining royalties in licensing agreements when evaluating assets during takeover procedures (special searches)

PATENTABILITY SEARCH. Approximately 20 percent of patent applications are rejected during prosecution because of the existence of documents anticipating the subject matter. It is recommended that a prior art search be made in every case before a patent application is prepared because it is much less expensive to conduct a patentability search than to prepare a patent application. If the search produces prior art which would probably preclude all worthwhile claims, one is spared the expense of preparing a patent application. Even if the art found is not anticipatory, the references are useful in preparing the application.

STATE OF THE ART SEARCH. A state of the art search is an in-depth search designed to give one an overview of the prior art in a general field. For an individual with an interest in a particular project, the search might involve scanning a few hundred United States patent documents. For a company establishing long term manufacturing policy for a new product line, the search would include patent documentation from several countries involving several thousand documents.

INFRINGEMENT SEARCH. An infringement search examines the claims of unexpired patents to determine if any unexpired patent would be infringed by a proposed activity. Since patent rights extend only in the country in which a patent is granted, infringement searches are limited to documents issued in the country of the proposed activity. For example, if one proposes to market a particular product in the United States, he would search unexpired United States patents issued in the last seventeen years.

For a defendant in an infringement action, it is important to know the identity of documents which can lead to the alleged infringed patent being declared invalid. In these cases, it is advantageous to perform a search in preparing for the case.

VALIDITY SEARCH. A validity search is an intensive search which usually includes foreign patent documents and research publications as well as United States patents. Here one is looking for a document that either anticipates or makes obvious another issued patent. On the other hand, if one is a defendant in a validity action, it is important to know the identity of documents which the opposition might present.

SPECIAL SEARCHES. There are a number of situations in which companies need extensive searches conducted to provide them with information for management decisions. For example, during takeovers, management needs to estimate the value of the patents held by the firm being taken over because future profits of the firm are related to its patent position. In a similar manner, this type of information is valuable in determining royalties in a licensing agreement.

DISTINGUISHING PATENT DOCUMENTS

With proper classification, a patent document dealing with any aspect of technology can be identified and retrieved. The purpose of a classification system is to provide an orderly arrangement of patent documents to facilitate access to the information contained within the system and to provide a base for dissemination of the information to all users of the system. The two major classifications systems are the United States Patent Classification System used by the United States, and the International Patent Classification System used by most other major patent countries.

United States patents are classified by the Patent Office according to subject matter into about 305 classes and 73,000 subclasses. Using the *Index to Classification,* the *Manual of Classification,* and the *Classification Definitions,* one can identify classes and subclasses of subject matter of interest. After identifying the class and subclasses of interest, one can retrieve the documents in these subdivisions for reading.

The United States Patent Office maintains a classified file of United States patents, arranged by subject matter, at its public search room in Arlington,

Virginia. At the public search room, the searcher can remove the selected subclass bundles of patents from the open stacks, take them to a desk, and read the patent documents, which are arranged in chronological order. By using the classification system, one can retrieve a file of several hundred patents in a specified subject area from the total base of five million plus documents.

Also at this location, the Patent Office maintains a scientific library for public use. This library maintains a collection of scientific books, journals, and indexes to the scientific literature. In addition, the library contains journals of foreign patent offices with many foreign patent copies. Except for the examiners' search files, foreign patents are not filed by subject matter, but are arranged in numerical order.

An additional classified patent file is maintained at various locations throughout the Patent Office complex for use by the patent examiners. In addition to the subclass patents, these files often contain unofficial subclasses, foreign patents, and some literature references classified by subject matter. The examiners' search files sometimes have "digests" of subject matter. These materials can be very useful in searching, and with permission, members of the public can use the examiners' search rooms as later discussed.

Unfortunately, the public search room is geographically inconvenient for regular use by most people. Certain libraries located throughout the United States are designated as "depository libraries" and maintain a United States patent collection and receive current issues of patents. These libraries are open to the public and maintain a staff to assist in a patent search.

To follow issuing patents, there are a number of abstracting services available. The *Gazette* is the official abstracting publication of the Patent Office, and it issues weekly. The *Gazette* provides a list of claims (usually the main claim) and a drawing of all patents issuing that week. By following the *Gazette,* one can keep abreast with weekly patent issues within a class and subclass.

Considering the number of patent documents published each week, over 20,000 worldwide, one can readily appreciate the need for abstracting services to prescreen, classify, and make available abstracts of these documents. There are a number of such companies available, and prominent among these are Chemical Abstracts, Derwent Patent Information, IFI/Plenum Data Company, and Pergamon Information. The services of these and other similar companies are discussed in a later chapter.

Most foreign countries classify their patent documents by the International Patent Classification System. Unfortunately, these countries do not in addition classify their patent documents by the United States Patent Classification System. The United States, however, uses the United States Patent Classification System for its classified patent files and uses the International Classification System for foreign dissemination.

Since foreign patent offices do not classify documents by the United States

Patent Classification System, the Classification Division must do so before these documents can be placed in the United States classified files. This is a tremendous task, and most non–English language documents have not been classified for the United States classified files. Foreign documents that have been classified by the United States system are filed in the examiners' search file, but not in the classified files in the public search room. While the examiners' search files do contain some foreign patents and publications associated with each subclass, this collection is not nearly complete.

It is sometimes necessary to search foreign patents and publications in addition to United States references as in the case of invalidity, infringement searches, and state-of-the-art searches. In these situations, it may be necessary to have a foreign search conducted by a foreign search facility. There are two European search facilities available to the public that will undertake a confidential search upon request without filing a patent application. These two facilities are the European Patent Office in Germany and the Swedish Patent Office at Stockholm. Both facilities will accept search requests from anyone, conduct an international search, and report the results in English. While these searches are expensive, they are quite good.

To keep track of recently published patent documents from foreign countries, one can follow the *Gazette* published by each office or one can subscribe to a commercial service which prescreens and abstracts issuing documents of different countries. Essentially all countries that publish patent documents in addition publish some type of gazette abstracting these documents. These gazettes are offered by subscription or can be read in a library. Copies of gazettes for most countries are available at the United States Patent Office. To use the gazettes effectively, however, one must be familiar with the International Patent Classification System.

As stated above, several companies prescreen, classify, and publish in English abstracts of recently published patent documents from foreign countries, including Chemical Abstracts, Derwent Patent Information Services, and the International Patent Documentation Center.

PATENT EQUIVALENTS

Many companies file patent applications for the same invention in more than one country. Such patent equivalents are commonly referred to as "patent families." Patent specifications within a family are usually, although not necessarily, similar in technical content. Since filing and prosecuting foreign applications is quite expensive, a patent family is usually a good indication that there is commercial interest in the technology. Identifying patent families can be useful for the following purposes:

1. It often provides an English-language equivalent of a foreign-language patent, thereby obviating the need for translations.

2. It indicates that a company has commercial interest in the inventive concept.

3. It indicates in which countries the patentee has patent protection.

The most effective method for identifying patent equivalents or families is through a patent concordance. Several companies publish such concordances, which are available to the public at libraries or by subscription.

Probably the most comprehensive concordance system is that offered by the International Patent Documentation Center (INPADOC) located in Vienna, Austria. For 21 countries, coverage goes back to 1968; in 1973 coverage was extended to 48 countries. This system covers a wide range of subject matter, but, it is limited in its time coverage. This service is marketed in the United States by Pergamon Information Company.

Other companies involved in this area are Derwent Publications Ltd. and Chemical Abstracts. These services are discussed in a later chapter. Suffice it to say at this point that a major disadvantage of the Derwent concordance is that it only goes back until about 1970 in most cases. *Chemical Abstracts* is limited to chemical and chemical engineering patents. Its coverage goes back to 1962 and is available at many libraries.

OBTAINING PATENT COPIES

Although patents are not subject to copyright laws, patent copies are often more difficult to obtain than research articles and other documents. Because patents are not as widely distributed as are many technical journals, books, etc. In addition, writing to an inventor for patent copies is not a fruitful approach since he probably does not have extra copies. There are, however, other sources by which one can obtain copies with a little more effort.

The most complete patent library in the United States is located at the United States Patent Office in Arlington, Virginia. The Patent Office maintains at this location a public search room with filed patent copies of all issued United States patents as well as copies of many foreign patent documents which can be photocopied. The public search files of United States patents are classified according to subject matter, which makes this a good location for searching these documents and for making copies.

Public and university libraries also maintain numerical sets of United States patents. Certain libraries throughout the United States are designated "patent depository libraries" and maintain a patent collection. These are open for public use and maintain photocopy equipment. Many industrial and research company libraries have extensive files of patents in their area of interest on microfilm or in printed form. Recently certain commercial companies began marketing microfiche of patents available to subscribers.

One can obtain patent copies from governmental patent offices or from private companies specializing in marketing patent copies. United States patents can be obtained by sending three dollars to the Commissioner of Patents and Trademarks, Washington, D.C. 20231. When ordering, identify the patent by patent number or by the name of the inventor and the approximate issue date. Foreign patent copies can be purchased from foreign patent offices. Addresses of foreign patent offices, along with prices for patent copies, are listed in a later chapter.

Although patent offices charge relatively low prices for patent copies, as a general rule they are very slow in delivery. As a result, many companies now specialize in patent procurement. While these companies are slightly more expensive, their service is more personalized and much faster.

Chapter 12

MAJOR PATENT OFFICES

INTRODUCTION

Over one hundred countries publish patent documents, and the number published annually is over one million. This chapter discusses the major industrial property offices of the world, including the United States Patent and Trademark Office, the European Patent Office, and the Patent Office of the Japanese government. Fifteen countries are associated with the European Patent Office. If the technology is important, it is likely that an application will be filed in at least one of these offices and probably in all three. This chapter discusses the basic features of these patent offices; details on how to obtain information from these offices are discussed in later chapters.

THE UNITED STATES PATENT AND TRADEMARK OFFICE

The United States Patent and Trademark Office was discussed in detail in Chapter 1. In addition to examining applications, the Patent Office publishes issued patents and other publications relating to patents. Listed below are some of these publications:

OFFICIAL GAZETTE. The *Gazette* is the official journal of the Patent Office. It is published weekly on Tuesday. It contains a selected claim and drawing of each patent issuing that week. These are arranged according to the United States Patent Classification System. The *Gazette* is sold by subscription by the superintendent of documents.

MANUAL OF CLASSIFICATION. Patents are classified by the Patent Office into a class and subclass according to their utility. This manual is a loose-leaf book containing a list of all classes and subclasses of this classification system. Classes and subclasses are constantly being reorganized, and substitute pages are issued from time to time. Sold by the superintendent of documents.

CLASSIFICATION DEFINITIONS. These definitions define in detail the subject matter assigned to each subclass. These are sold by subclass by the Patent Office.

WEEKLY CLASS SHEET. This is a list showing the classification of each patent in the *Gazette*. Sold by subscription by the superintendent of documents.

PATENT LAWS. This is a list of patent laws in force and is sold by the superintendent of documents.

TITLE 37 CODE OF FEDERAL REGULATIONS. This includes the rules of practice for patents, trademarks, and copyrights. Sold by the superintendent of documents.

DIRECTORY OF REGISTERED PATENT ATTORNEYS AND AGENTS. This is a geographical listing of persons registered to practice before the Patent Office. Sold by the superintendent of documents.

MANUAL OF PATENT EXAMINING PROCEDURE. This is a manual of reference work for examining practice and procedure for the Patent Office. Sold by the superintendent of documents.

THE EUROPEAN PATENT OFFICE

The European Patent Convention was established in 1970 for the purpose of establishing an office to grant European patents. This convention established the European Patent Office for centralized filing, searching, and examining applications for member countries. If the application matures, a bundle of individual European patents are granted, one for each country designated by the applicant. Member countries of the European Patent Office include the following:

Austria	Liechtenstein
Belgium	Luxembourg
Denmark	Monaco
France	Netherlands
Germany	Spain
Great Britain	Sweden
Greece	Switzerland
Italy	

The European Patent Office, established by the European Patent Convention, came into being in November 1977. It is headquartered in Munich, Germany. On January 1, 1978, the International Patents Institute ceased to exist as an independent organization and was incorporated into the European Patent Office. The function of the International Patents Institute had previously been to perform state of the art searches on patent applications filed with the patent offices of its member states.

The European Patent Office has two main functions: to perform formal searches and examinations and issue patents for its member countries and to function as an International Searching Authority providing state of the art searches for the public. The latter function provides an important method by which the public can obtain patent information.

PATENT OFFICE OF THE JAPANESE GOVERNMENT

The Patent Office of the Japanese Government (POJG) issues two types of industrial property documents: the patent and the utility model. As a general rule, the patent is concerned with both tangible inventions (devices, machines, etc.) and intangible inventions (methods, processes, etc.); the utility model is concerned only with tangible inventions. Therefore when searching methods, processes, etc., one needs to search only patents; when searching machines, devices, etc., one needs to search both patents and utility models. The patent and utility model do not differ in rights, but do differ in their terms: 15–20 years for the patent and 10–15 years for the utility model.

All patent and utility model applications are published before examination, usually 18 months after the filing date. These are then examined by filing order; some are rejected, some are published for opposition. There are four kinds of documents issued by the POJG.

1. KOKAI patents (unexamined)
2. KOKOKU patents (examined)
3. KOKAI utility models (unexamined)
4. KOKOKU utility models (examined)

Prior to 1975, Japanese patent documents were classified only by the Japanese Patent Classification System; from 1976 to 1979 documents are classified by both the Japanese Classification System and the International Patent Classification System; from 1980 documents are classified only by the International Patent Classification System.

TREATIES

Since technology and the utilization of technology are nonnational in character, it is advantageous for countries to have international agreements for the protection of patent rights. The United States has signed two treaties relating to patents and trademarks with other industrial countries as discussed below.

Paris Convention of Industrial Property

The United States has a treaty with 78 other industrial countries relating to the intellectual properties of the patent and trademark. This treaty is known as the Paris Convention for the Protection of Industrial Property and dates back to 1883. Essentially all industrialized countries are participating countries.

This treaty provides that each country guarantees to citizens of other participating countries the same rights in patent and trademark matters as it provides its own citizens. In addition, the treaty provides for the right of priority in cases of patents and trademarks. The foreign priority right provides that when an applicant has filed an earlier foreign patent application for the same invention disclosed in a United States application, he may claim the earlier filing date of the foreign application as the effective filing date of his United States application. Likewise, participating countries afford foreign priority to earlier filed United States applications when filing in that country. To obtain foreign priority, the application must be filed within twelve months (six months in case of design patents) after filing the foreign application upon which the application relies for priority.

When a United States application claiming foreign priority is properly filed, it is regarded as if the application had been filed on the same day as the foreign application on which it relies. Therefore certain acts such as publication, public use, and public sale of the subject matter committed between the foreign filing date and the United States filing date cannot be used to anticipate the claims of the application. The application also has priority over other applications for the same invention filed during this interval.

Patent Cooperation Treaty

During recent years there has been a move toward multinational filing and prosecuting of patent applications. This simplifies and decreases the expense of foreign filings. The Patent Cooperation Treaty is an agreement between certain nations for the purpose of simplifying the procedures for residents of a member country filing in another member country. The treaty came into effect in 1978. More than twenty countries adhere to the treaty, including the United States, Japan, Australia, and most European countries.

The purpose of the treaty is to facilitate the filing of applications for patents on the same invention in member countries by providing for centralized filing and a standard application format. An international search is conducted, and copies of the international application and search report are then distributed to each member country designated by the applicant. Examination of the application is carried out by the national patent office of the designated countries.

Chapter 13

THE UNITED STATES PATENT CLASSIFICATION SYSTEM

INTRODUCTION

Approximately one hundred countries publish patent documents, and in recent years the number of patent documents published annually has been over one million.

These documents deal with practically every aspect of technology and contain an immense amount of prior art information. For these documents to be of value as prior art, there must be classification systems allowing retrieval of the documents of interest.

With proper classification, a patent document dealing with any particular aspect of technology can be identified and retrieved. A classification system should meet the following objectives:

(1) To provide an orderly arrangement of patent documents to facilitate access to the information contained within the system

(2) To provide a basis for investigating the state of the art in any field of technology; this would be necessary for examiners determining patentability of subject matter in applications and for others performing novelty searches, validity searches, and infringement searches

(3) To provide a basis for dissemination of the information to all users of the patent information system

The two major classification systems are the United States Patent Classification System (USPCS) and the International Patent Classification System (IPCS). The United States uses the USPCS, and most other major patent countries use the IPCS. The United States Patent Office asserts that the USPCS has several advantages over the IPCS and will probably continue to use this system indefinitely. This author agrees that it does have certain advantages.

This chapter is concerned with the details of the United States Patent

Classification System; the International Patent Classification System is discussed in the next chapter. To fully utilize the United States patent collection as a prior art source, it is necessary to understand the structure and principles of the classification system. An understanding of the basis of this system allows one to effectively identify and retrieve documents in a specific subject area.

HISTORY

The USPCS was initiated in the middle of the nineteenth century. In 1830, Congress authorized the publication of a subject-oriented patent listing, which contained 22 classes. From about 1840 to 1870, the USPCS underwent several revisions. It was constantly being adapted to keep pace with the rapidly growing patent system. During this period, the British and French classification systems had a large influence on the United States system. The number of classes increased to 26, and subdivisions were included in many classes.

In 1872 the USPCS was revised by an expanded publication setting forth 165 classes; these classes are the framework upon which the present system is built. In 1898, Congress established the Classification Division of the Patent Office. Since then the classes have constantly been revised and modified by this division as the need arises.

In 1966 the Patent Office issued a publication entitled *Development and Use of Patent Classification Systems*, which supersedes all previous publications. This publication sets forth a new format whereby new classes and subclasses are to be established each year as technology changes. The USPCS is therefore a dynamic system. About four percent of the Patent Office search files are affected by this process at any given time. At times an entire class or subclass is removed from the public search room for reclassification. During this period, copies of these classes must be searched in the examiners' search files.

Since 1974 the Patent Office has had a computer data base containing an inventory of United States patents in each class and subclass. From this data base, the Classification Division can follow changing technology and adjust reclassification accordingly. Public use of this data base is discussed in a later section.

STRUCTURE

The United States patent classification system is an arrangement of subject matter into classes and subclasses to facilitate the retrieval of this information when desired. In theory, the USPCS is an arrangement of all scientific and technical information encompassed by all prior art, that is, prior art claimed and disclosed in domestic patents, prior art disclosed in foreign patents, and prior art disclosed in all other nonpatent printed publications. The primary purpose of the system is to make this information available to the patent examiners for examining patent applications; the secondary purpose of this system is to make this prior art information available to inventors for determining if their discoveries meet the novelty and nonobviousness requirements for patentability and to any others with an interest in ideas, information, or facts.

In particular, this information aids those conducting state of the art searches, validity searches, infringement searches, etc.

Classes and Subclasses

In the USPCS, science and technology is divided into about 305 main categories called classes. Each class is a general category covering related subject matter or types of inventions, for example, Class 100, "Presses"; Class 131, "Tobacco"; Class 364, "Electrical Computers and Data Processing Systems."

The system further breaks down each class into smaller categories called subclasses. The primary divisions within each class are called mainline subclasses. The mainline subclasses are further broken down into other subclasses. There are approximately 73,000 mainline and other subclasses.

Each class and subclass is titled and numbered. The *Manual of Classification*, a Patent Office publication, lists the numbers and titles of all classes and subclasses. A sample is illustrated, showing a partial schedule of Class 364, "Electrical Computer and Data Processing Systems," subclasses 130–413, taken from the *Manual of Classification*.

The position of the subclass within each class is important. Mainline subclasses are always in full capitals, are listed farthest to the left, and are not indented. In the accompanying example, subclass 130, DATA PROCESSING CONTROL SYSTEMS, and subclass 400, APPLICATIONS, are mainline subclasses.

The mainline subclasses have indented under them a number of other subclasses. Referring to the example, under subclass 130, there are a number of subclasses indented one level (one dot), including subclasses 131, 137, 138, 140, and 148. Subclasses indented to the same level are called "coordinate subclasses."

The above subclasses are further broken down by subclasses indented two levels (two dots). For example, subclass 131, "Plural processors," is further

130	DATA PROCESSING CONTROL SYSTEMS, METHODS OR APPARATUS	181	..Manual/automatic
131	.Plural processors	182	..Fine/coarse
132	..Master-slave	183	.With specific error signal generation (e.g., up/down counter)
133	..Parallel		
134	...Shared memory	184	.With protection or reliability feature
135	..Hybrid types (analog, digital)	185	..Warning or alarm
136	..Including sequence or logic processor	186	..Self-test
137	.Cascade control	187	..Backup/standby
138	.Supervisory control	188	.With operator control interface (e.g., control/display console)
139	..Of analog controllers		
140	.Sequential or selective	189	..Keyboard
141	..State of condition or parameter (e.g., on/off)	190	..Positional (e.g., joystick)
		191	.With preparation of program
142	...Position responsive	192	..Editing/modifying
143	...Time responsive (duration)	193	..Playback
144	....With display	194	.With specific algorithm
145	....Clock-calendar (e.g., time of day)	400	APPLICATIONS
146	..Operator interface (e.g., display with controls)	401	.Business practice and management
		402	..Operations research
147	...Specific programming (e.g., relay or ladder logic)	403	..Inventory
		404	...With cash register
148	.Optimization or adaptive control	405	..Cash register
149	..With model	406	..Accounting
150	...Comparison with model (e.g., model reference)	407	..Reservations
		408	..Finance (e.g., securities, commodities)
151	...With adjustment of model (e.g., update)		
		409	.Government activities (e.g., voting, law enforcement)
152	..Specific criteria of system performance		
		410	.Games and amusements
153	...Constraints or limits (e.g., max/min)	411	..Scoring
154	....Variable	412	..Wagering
155	....Bidirectional (e.g., oscillatory)	413.01	.Life sciences
156	....Economic (e.g., cost)	413.02	..Patient monitoring or diagnostics
157	..Gain (e.g., tuning)	413.03	...Vital signs (e.g., respiration, temperature, blood pressure, pulse)
158	..With perturbation		
159	...Test signal		
160	..Plural modes	413.04	....Physiological conditioning system
161	...Proportional-integral (P-I)	413.05	...Wave or rhythm
162	...Proportional-integral-derivative (P-I-D)	413.06	....Electrocardiogram
		413.07	...Blood
163	...Proportional-derivative (P-D)	413.08	....Blood cell analysis
164	..Feed-forward (e.g., predictive)	413.09	....Blood chemistry (e.g., oxygen level)
165	...Combined with feedback	413.1	...Cellular composition or activity
166	..Rate control	413.11	...Body chemistry (e.g., urine analysis)
167.01	.Digital positioning (other than machine tool)	413.12	...Fertility cycle
		413.13	..Medical imaging
172	.Plural variables	413.14	...Computed tomography using X-ray
173	..Ratio	413.15	....Particular data acquisition technique
174	..Positional (e.g., velocity, acceleration)		
175	...Positional with nonpositional	413.16	....Particular projection data set creation technique
176	.Specific compensation or stabilization feature		
		413.17	.....Weighting factors
177	..Lag (e.g., deadtime)	413.18	.....Interpolated or extrapolated data
178	.Sampled data system	413.19	....Particular image reconstruction technique
179	..Variable rate		
180	.Multiple modes (e.g., digital/analog)	413.2	.....Fourier transformation
		413.21	.....Convolution or back projection
		413.22	....Image display

Partial schedule of Class 364, "Electrical Computer and Data Processing Systems," subclasses 130–143, from the *Manual of Classification*.

broken down into subclasses 132–137. Subclass 133 is further broken down into subclass 134.

The degree of indentation relates to the concept of "superiority" which is discussed in section IV. In general, the subject matter of a subclass includes that indicated by not only its own title, but also by the title of each subclass under which it is indented.

Classification Definitions

Each subclass has a title which generally defines the subject matter in the subclass. The subclass titles are rather brief in the *Manual of Classification.* Each class and subclass is further defined by the "classification definitions." The classification definitions are statements of the scope of subject matter encompassed by each class and subclass along with search notes which direct searchers to related subject matter in other classes and subclasses.

In the USPCS, each class and subclass must be defined. The title of each class or subclass must be explained in a detailed statement setting forth the metes and bounds of the area of subject matter for each class and subclass. A class and subclass definition must include a description of the subject matter encompassed by the class or subclass, and it may also include explanatory and search notes.

In certain classes and subclasses, to supplement or take the place of cross referencing (later discussed), search notes are included. These give directions and suggestions for further search, setting out the relationship and lines of distinction between classes and subclasses. Search notes usually indicate other classes or subclasses directed to related subject matter. Search notes also indicate classes or subclasses directed to subject matter constituting either a combination or subcombination of the class or subclass in which the note is written.

Cross-Referencing

The USPCS is an obligatory-type classification system based on the claims; that is, all United States patents must have one classification symbol and this is called the "original classification." A document can have only one original classification symbol. According to the rules of the system, the original classification is based on the most comprehensive claim in the patent. This claim is referred to as the controlling claim.

The USPCS also has additional classifications which are called "cross-reference classifications." These are based on the fact that nearly every patent document discloses subject matter other than that in the most comprehensive claim and subject matter that is classifiable in a different class or subclass than that which provides for the subject matter of the controlling claim. In the USPCS such different subject matter is provided for by the assignment of one or more cross-reference copies. This subject matter falls into one of two categories: subject matter which is separately claimed in a claim other than the controlling claim or subject matter which is disclosed but not claimed in a claim other than the controlling claim.

If the disclosed subject matter is claimed, it is obligatory for the classifiers to cross-reference the patent to the subclass or subclasses providing for the subject matter of such other claims, unless search notes are provided which would lead a searcher to the subclass to which the patent is assigned on the basis of the most comprehensive claim.

Any disclosure in a patent which is disclosed but not claimed may be cross-referenced into any part of the classification system at the discretion of the classifier. The following criteria are considered for such cross-referencing: the disclosure must in the judgment of the classifier be novel, and the disclosure must be of sufficient detail and clarity to be useful as a reference. No cross-reference is made when a search note is appended to the definition of the subclass eligible to receive the cross-reference, indicating that the subclass containing the original copy of the patent must be searched.

Unofficial Subclasses and Digests

Over the years, patent examiners have created unofficial subclasses and digests to facilitate searches within the art under the jurisdiction. These unofficial subclasses and digests are listed and designated in the *Manual of Classification*. In order to distinguish between unofficial subclasses, digests, and official subclasses, certain distinctions are made.

Unofficial subclasses are groups of patents selected by an examiner. In the *Manual of Classification*, these are made an indented subclass under the official subclass with an alpha designation following the official numeric designation. Since the original subclass then does not have all the patents officially classified therein, it is given the alpha designation "R" (indicating Residual). It is noted that the numeric subclass equates to the residual subclass plus the unofficial subclasses indented thereunder.

The unofficial subclasses are double indented under the subclass from which they were formed. This indicates that rules of superiority (later discussed) do not apply to these subclass groupings. There are no definitions for the unofficial subclasses. The unofficial subclasses are filed separately in the examiners' search files. The public search room file contains only those patents listed according to the official (numeric) classification, however; unofficial subclasses are not separately filed in the public search room. These collections can be helpful in certain search situations and can be searched in the examiners' search files.

Digests are collections of copies of patents based on a concept which relates to a class but not to any particular subclass of that class. These are established by examiners to help in searching their art. An original classification of a patent cannot include a digest because digests are outside of the official classification arrangement. The digests are listed in numeric sequence at the end of each class schedule in the *Manual of Classification* and can be searched in the examiners' search files.

PRINCIPLES OF CLASSIFICATION

The USPCS is designed primarily to supply prior art information to the patent examiner to determine patentability, novelty, and interference of pending applications. The principles of this classification system differ somewhat from the classification system used in most libraries. Most library classification systems are designed primarily for finding physical objects. Patents, on the other hand, are concerned with ideals, information, and facts, and a system suitable for storage and retrieval of all of these is of utmost importance in a patent classification system.

While the USPCS system is designed primarily to supply prior art information to patent examiners, it is a highly organized system which can be readily understood by others. Once understood, the United States patent collection becomes more readily available as a prior art source and can be used more easily for conducting patentability searches, state-of-the-art searches, validity searches, infringement searches, etc.

Dupacs

A *Patent Office publication, Development and Use of the Patent Classification System* (DUPACS), Library of Congress No. 6562235, sets forth the principles and basis of the USPCS. This work was published in 1966 and serves as the guideline for the Classification Department in the Patent Office. It supersedes all previous classification publications.

This publication contains vital information for understanding and using the United States patent collection as a prior art source. This publication consists of eight chapters listed below:

    1. An Introduction to the U.S. Patent Classification System
    2. Bases of Classification
    3. Analysis of Scientific and Technical Subject Matter and Subdivision of Such Subject Matter
    4. Creating a Single Class
    5. Title, Definition, Notes and Crossreferencing
    6. Classification Project: Practice and Procedure
    7. How to Use the U.S. Patent Classification System
    8. Administration of Existing Classes

For this discussion, Chapters 1, 2, and 7 are of interest since they relate to the bases and use of the classification system. The remaining chapters are designed primarily for the patent classifiers, but may be of interest to a user of the system after he has developed a basic understanding of the system. If the user of the system understands the principles by which documents are assigned to the system, it follows that it will be easier to retrieve documents from the system.

Basis of Classification

Since all patentable subject matter is created for its utility, utility is the characteristic that has been elected as the primary basis for classification. However, since most inventions contribute to numerous utilities, another concept has been built into the basis of classification, the concept of "proximate function." Under this concept, devices or processes that inherently achieve similar results are classified together regardless of their stated utility. Arts or instruments having like functions, producing like products, or achieving like effects are classified together. For example, all cutting machines are classified together regardless of whether they cut wood, cheese, metal, or meat because they share the proximate function of utility.

The basis of classification is summarized in DUPACS, Sects. 1–5, cited below.

> 1. **Utility as a Basis of Classification**   The principle basis for classifying the useful arts in the U.S.Patent Classification System is utility, that is, the function of a process or means or the effect or product produced by such process or means. Utility as a basis of classification must be taken in the sense of direct, proximate, or necessary function, effect or product rather than remote or accidental use or application as in industries or trades. Applying proximate function, effect, or product as a basis of classification will result in collecting together similar processes or means that achieve similar results by the application of similar natural laws.
>
> 2. **Proximate Function as a Basis of Classification**   Proximate function as a basis of classification is generally applied to processes or means for performing general operations in which 2 single causative characteristics can be identified and which requires essentially a single unitary act.
>
> 3. **Proximate Effect or Product as a Basis of Classification**   Effect or product as a basis of classification is generally applied to complex special results of a process or means requiring successive manipulations involving plural acts.
>
> 4. **Structure as a Basis of Classification**   Structural features such as the configuration or physical make-up of a means may be used as a basis of classification only when the subject matter to be classified is so simple as to have no clear functional characteristics, but can only be distinguished from other subject matter by its structural features. This situation rarely arises with respect to the creation of a large group or class in the system, but frequently occurs with respect to subdivisions within a large group or class. As between a classification system based upon structure and one based upon proximate function, effect, or product, the choice is for the latter in all situations in which it can be applied.

5. **Basis of Classification Applicable to Chemical Compounds and Mixtures or Compositions** A chemical compound should be classified on its structure, that is, on the basis of its chemical constitution, regardless of the utility thereof. Mixtures or compositions, at least in the larger groupings, are generally collected on the basis of the disclosed utility for the particular material.

Section 1 sets forth the basis for classifying the useful arts in the USPCS. Early in the history of the USPCS, patent classification was largely industry-oriented; inventions were classified according to the branch of industry or art to which they were relevant. As the patent system progressed, it became evident that this approach was not completely meeting the needs of the United States patent system. The system was restructured and the concept of "proximate function" was adopted as the basis of classification. Under this concept, processes and devices that achieve similar results by application of similar natural laws are classified together, regardless of their stated utility.

Section 4 relates to structure as a basis of classification. For processes, devices, and articles of manufacture, structural features such as configuration or physical makeup may be used as a basis of classification only when the subject matter of the disclosure is so simple no clear functional characteristics exist. Therefore, for these types of inventions, when the choice for basis of classification is between structure and proximate function, effect, or product, the choice is always proximate function, effect, or product if these concepts can be applied.

Section 5 relates to the basis of classification for chemical compounds and compositions. Chemical compounds are classified on the basis of their structure, that is, on the basis of their chemical constitution regardless of their stated utility. On the other hand, mixtures and compositions are classified on the basis of their disclosed utility.

The USPCS is an obligatory-type classification whereby classification is based on the claimed subject matter. This is set forth in DUPACS, Sects. 6–8, cited below.

6. **Analysis as a Prerequisite to System Development** The U.S. Patent Classification System is created by analyzing the disclosures of U.S. patents and then creating classes (including the schedule of subclasses within each class) by grouping together like subject matter as represented in the disclosures of such patents.

7. **Patents Grouped by Claimed Disclosure** Inasmuch as nearly every U.S. patent contains disclosure that is claimed and also disclosure that is not claimed, the general principle is that a classification system is created and a patent shall be assigned therein on the basis of that portion of the disclosure covered by the claims rather than on a portion of the disclosure that is not claimed. A disclosure that is not claimed is one that may form an element or step of a claimed combination as well as a disclosure not referred to in any claim.

8. Patents Diagnosed by Most Comprehensive Claim   The totality of a claimed disclosure must be selected, whenever possible, in creating a classification system and determining the appropriate class to which a patent is assigned, but a mere difference in the scope or breadth of claims should not make a difference in assignment.

The claim is the part of the patent which sets forth what is new and thus protected by the patent rights. According to the rules of the USPCS, classification among classes is based on the most comprehensive claim in the patent. The most comprehensive claim is the claim that encompasses more of the disclosed subject matter than any other claim. The original classification is based on the most comprehensive claim in the patent. A patent document can have only one original classification.

Under rare circumstances there are exceptions to the claimed disclosure principle for assigning patents to a specific class. These exceptions are set forth in DUPACS, Sect. 9 and 10, cited below.

9. Exceptions to Claimed Disclosure Principle for Assigning Patents to Specific Class   The following situations are exceptions to the principle that a system is created and the patents assigned therein on the claimed disclosure of U.S. patents. When these exceptions are applied, it should be clearly stated in the class definitions of the classes involved.

A. Old Combination With Specific Subcombination   Where a patent claims a specific subcombination in combination with some other broadly recited subcombination, the combination and subcombination being classified in different classes, there are exceptions to the general principle that a classification system is created and a patent is assigned on the basis of the claimed disclosure; that is, the patent may be assigned to the subcombination class when all the following conditions apply:

(1) A relatively large number of patents are involved.

(2) The combination is old as a matter of common knowledge.

(3) No new relationship between the subcombinations is set forth.

(4) The other subcombination is nominally claimed.

B. Article Defined by Material From Which It Is Made   A patent for an article of manufacture, claimed by name only and in which the claim is otherwise directed to a specific material of which the article is made, is generally assigned to a class providing for the material rather than a class providing for the article.

C. Process of Utilizing a Composition   A patent claiming a process of utilizing a specifically defined composition may be assigned to the composition class where the process steps are nominally recited and the composition class provides specifically for compositions having that use.

10. Exception to Claimed Disclosure Principle for Patent Assignment Between Subcombination Subclass and Indented Combination Subclass   Where a parent subclass has indented thereunder a combination subclass which includes as a subcombination thereof the subject matter of the parent subclass, a patent disclosing the subject matter of the combination subclass but claiming only the subject matter of the subcombination subclass is assigned to the indented combination subclass.

To summarize, the USPCS is an obligatory classification system based on the claims. All United States patents must have one classification symbol called the "original classification," and a patent document can have only one original classification symbol. Documents can have additional classification symbols called "cross-reference classifications," however. The cross-reference classifications arise because nearly every patent discloses subject matter other than that in the most comprehensive claim. This is more fully developed in the next section.

Superiority of Classes and Subclasses

Another concept essential to an understanding of the classification system is superiority. Superiority is a set of principles set forth by the Patent Office for establishing priority for assignment of subject matter to classes and among subclasses within a class. While these principles were established to aid in the ongoing assignment by the Patent Office of documents into the system for storage, a basic understanding of these principles is essential for searching in the system for retrieval of prior art documents.

The process of assigning an application or patent in the system is in many respects similar to the process used for retrieving documents from the system. In both cases, it is necessary to have an understanding of the subject matter involved. This section is concerned with the rules for assigning documents to the system and in particular with how these rules are applied when selecting the proper class and subclass for retrieving documents of specific subject matter.

SUPERIORITY OF CLASSES. As previously discussed, the USPCS is by nature an obligatory type of system whereby a patent document must be designated by one "original classification" denoted by one class and subclass number; there can be additional cross-reference classifications. The Patent Office Rules from DUPACs, Sect. 720, for determining superiority among classes are cited below.

SUPERIORITY AMONG CLASSES

A. With respect to an application or patent directed to one claimed disclosure assignment is to the class that is the locus of the prior art for the same subject matter. The identity of the proper class is established through study of class definitions and notes of classes suggested by the Index to Classification or lists of classes or by personal knowledge of the location of the prior art.

B. With respect to an application or patent including claimed disclosures to diverse inventions, the principles listed below must be considered and applied, if appropriate, stepwise, in the order listed to select the single disclosure that will control assignment as in A above:

(1) Most comprehensive claimed disclosure governs.

(2) Order of superiority of statutory categories of subject matter.

    (a) Process (of using product b, e.g., using a fuel or radio transmitter)

    (b) Product (of manufacture, e.g., a fuel or radio transmitter)

    (c) Process (of making product b)

    (d) Apparatus (to perform c or to make b, e.g., machine, tool, etc.)

    (e) Material (used in c to make b)

(3) When, and only when, principles 1 and 2, given above, fail to solve the question of the controlling class, the relative superiority of types of subject matter as shown by the following list is used:

    (a) Subject matter relating to maintenance or preservation of life is superior to subject matter itemized in b-d below.

    (b) Chemical subject matter is superior to electrical or mechanical subject matter.

    (c) Electrical subject matter is superior to mechanical subject matter.

    (d) Dynamic subject matter (i.e., relating to moving things or combinations of relatively movable parts) is superior to static subject matter (i.e., stationary things or of parts nonmovable related).

The above guideline distinguishes between documents directed to one claimed disclosure and documents directed to more than one claimed disclosure. Documents with one claimed disclosure are assigned to the class of prior art for the same subject matter; documents with more than one claimed disclosure are assigned to a class according to a series of principles set forth.

Subsection A relates to documents with one claimed disclosure. A majority of patents and applications have only one claimed disclosure defining a simple invention. In these cases, the invention is defined as a simple whole product, a process, or machine. This type of document presents no problem in determining which claimed invention should be chosen for making assignment. The class is simply selected which provides for the subject matter of the claimed disclosure. The definitions and notes set forth superiority among the classes which provide for related subject matter, either by positive statements of the distinctions between certain classes or by stating the location of other related subject matter.

Subsection B relates to documents with more than one claimed disclosure. Some patents or applications include diverse types of claimed disclosures, for example, combinations and subcombinations, products and processes, etc. Since the different claimed disclosures could be classified in different classes, this type of document presents a problem as to which of the claimed disclosures should control assignment of the document. In these cases, a determination of controlling subject matter is made, and assignment is based on that claimed disclosure as though it were the only disclosure in the patent or application.

The following principles are applied to determine which claimed disclosures control assignment. In essence, the claimed disclosure setting forth the most comprehensive combination controls assignment. For example, when a claim to a combination is compared to a claim to a subcombination, the combination claim controls assignment. By this principle, it is necessary to scan and compare all claims in order to select the controlling claim, that is, the claim directed to the most comprehensive combination when compared to other claimed disclosures.

When the plural claimed disclosures are of the same comprehensives and cannot be assigned by the above principle, superiority is determined by categories of subject matter. In patent law, inventions are divided into four statutory categories of inventions as set forth in Section 101 of the Patent Code. These categories include processes, machines, manufacture, or compositions of matter. These categories are defined as following:

PROCESS. An act, or series of acts, performed upon an object producing some change in the character, condition, or place of the object; mechanical as well as chemical processes are included.

MACHINE. A machine includes any mechanical device or combination of mechanical powers and devices which perform some result or effect.

MANUFACTURE. The term manufacture in patent law includes everything which is made by the art or industry of man not being a machine, composition of matter, or a design.

COMPOSITION. Any mixture of two or more substances; may be a mechanical mixture, or a product of a chemical reaction.

The following order of superiority is set forth for statutory categories of subject matter: process of using product, product of manufacture, process of making product, apparatus or machine, and material. The highest position in the order controls assignment.

When principles 1 and 2 discussed above fail to solve the question of controlling class, class superiority is determined by the type of subject matter disclosed in the document. The following order of superiority has been set forth: subject matter relating to maintenance of life has overall superiority, chemical subject matter is superior to electrical or mechanical subject matter electrical subject matter is superior to mechanical subject matter, and dynamic subject matter is superior to static subject matter.

Superiority Among Subclasses Within a Class

Once the proper class has been identified, it is necessary to select the proper subclass for searching by scanning the class schedule from the top to bottom subclass. The relative location of a subclass in the array of subclasses determines its superiority. The top (first) subclass has the highest priority, and the bottom (last) subclass has the lowest priority. DUPACS, Sect. 720, sets forth a summary of principles of superiority among subclasses of a class and is cited below.

Superiority Among Subclasses of a Single Class

Among subclasses within a class schedule, the first subclass reached, upon scanning coordinate subclasses from top to bottom, that provides for the claimed disclosure governs. As between coordinate subclasses each providing for a different characteristic, a claimed disclosure to a combination of the two is placed in the first appearing subclass in all instances where such combination is not provided for in some preceding subclass or in some other class and a cross-reference is assigned to the other. Assignment is to be carried into indented subclasses, if any, under the so determined coordinate subclass until the ultimate indented subclass is reached that provides for the claimed subject matter.

After identification of the class, the proper subclass is selected by scanning the schedule from top subclass to bottom subclass. On the initial downward scan, only first line subclasses are considered. Each first line subclass is evaluated on the way downward until a first line subclass is reached which provides part or all of the claimed disclosure of the search. The selected first line subclass, and its indented subclasses, provide the basic search schedule. If the first line subclass has coordinate subclasses indented under it, these indented subclasses should be scanned in a downward manner, carefully evaluating each individually. If an indented subclass itself has coordinate indented subclasses, this process is repeated until one has tuned in to the ultimate indented subclass containing subject matter of interest.

Additional principles that might be of value in determining subclasses of interest during a search are set forth in DUPACS, Sects. 12–15, cited below.

12. **Exhaustive Nature of Coordinate Subclasses: Combinations To Precede Subcombinations**   Coordinate subclasses must each be exhaustive of the classification characteristic for which the subclass title and definition provides. That is, no subsequent coordinate subclass nor any subclass indented thereunder should provide for the characteristic of an earlier appearing coordinate subclass. Thus, in coordinate relationship, combinations including a detail must precede subcombinations to the detail, per se. A subsequent subcombination subclass receives disclosed combinations—which in their entirety are provided for in a preceding subclass where only the subcombination is claimed; the disclosed combination is cross-referenced, if appropriate, to such preceding subclass.

13. **Indentation of Subclass**   A class schedule is arranged with certain subclasses appropriately indented. In a properly indented schedule, subclasses at the extreme left in a column of subclasses are the main variants (referred to as "first line subclasses") of the class. The titles and definitions of all these first line subclasses must be read with the title and definition of the class, as if indented one space to the right under the class title.

A subclass having indented subclasses under it represents a subject divided into variants. Such subclass also includes other variants not comprehended by the indented variants. If no genus subclass is provided for the concepts of several subclasses which are in fact variants of a genus, the several subclasses should be positioned in the same area of the schedule where possible, as though they were indented under the unprovided-for genus.

14. **Diverse Modes of Combining Similar Parts**   The classification system must recognize and provide for diverse modes of combining the same or similar parts or steps to obtain functionally (and possibly structurally) unrelated combinations.

15. **Relative Positions of Subclasses** The relative position of subclasses in a single class is determined by the following principles:

(1) Characteristics deemed more important for purposes of search generally should be provided for in subclasses that precede subclasses based on characteristics deemed less important. However, some subclasses of lesser importance may require precedence of position to avoid their loss from the schedule.

(2) Subclasses based upon effect or special use should precede those based upon function or general use.

(3) Subclasses which are directed to variants of a concept should either be indented under the subclass directed to such concept or precede the same, and should not form or be part of a subsequent coordinate subclass or group of subclasses.

(4) Subclasses directed to combinations of the basic subject matter of the class with means having a function or utility unnecessary for or in addition to the function or utility of the basic subject matter should precede subclasses devoted to such basic subject matter.

HOW TO USE

The Classification Division of the Patent Office publishes three publications which are essential in the effective use of the USPCS. These publications are:

1. The *Index to Classification*
2. The *Manual of Classification*
3. *Classification Bulletins and Subclass Definitions*

Copies of these publications are available in the public search room at the Patent Office and at various libraries throughout the country, or they may be purchased. The *Index* and *Manual* can be purchased from the Superintendent of Documents, U.S. Government Printing Office, Washington, D.C. 20402. *Subclass Definitions* can be purchased, class by class, from the Patent Office, Special Processing Section, Reclassification Branch, Washington, D.C. 20231. Since definitions are sold class by class, one can purchase definitions in classes of interest. Purchased this way, total cost is reasonable. A price list of the definitions can be obtained from the above address.

Index to Classification

The *Index to Classification* is an introductory key to the classification system and is useful in locating clues to an unknown field of search. The *Index*

INDEX TO CLASSIFICATION

	Class	Subclass
Axminster		
Carpet	139	399
Making apparatus	139	2
Azeotropes	203	50
Azides	423	
Inorganic metal azides	149	35
Explosive or thermic containing	149	35
Organic radical containing	260	349
Azimuth Instrument		
Ammunition	102	49.1
Horizontal and vertical angle	33	281
Horizontal angle	33	285
Optical	356	138
Solar locating	33	268
Telescope	350	8
Azines	544	1
Azines	564	249
→ Azo compounds	260	153
Heavy metal containing	260	146
Azo Compounds	260	144
Dye compositions containing	8	662
Several dyes	8	639
Textile printing	8	445
Azoles	548	100
Acridine nucleus containing	546	-26
Aluminum containing	548	101
Arsenic containing	548	102
Azo compounds	260	157
Heavy metal containing	260	147
Boron containing	548	110
Heavy metal containing	548	101
Phosphorus containing	548	111
Silicon containing	548	110

	Class	Subclass
Bacitracins	260	112.5R
Back		
Pad harness	54	66
Rest	297	452
Bed	5	70
Boat	9	7
Design	D 6	200
Scratcher	128	62R
Backband		
Harness	54	4
Backfire	261	DIG.6
Backfire Preventer	48	192
Backgrounds Photographic	354	291
Backing Dental		
Instrument	433	141
Making	29	160.6
Metalware shaping	72	54
Pliers	81	5.1R
Backlash Take up		
Between meshing gears	74	409
Sectional gear	74	440
Milling work feeds	409	146
Backpack	224	153L
Backrest	D 6	200
Backup Auto Lights	362	257
Bacon	426	645
Packaging	53	DIG.1
Preservation	426	332
Bacteria	435	
Fertilizer preparation with	71	6
Liquid purification by	210	601
Virus culture on	435	235
Bactericidal		
Compositions	424	

Sample from *Index to Classification*.

is an alphabetical listing of technical and common names of processes, machines, articles, composition of matter, etc., with a corresponding numerical citation to a class and subclass of the system in which pertinent prior art is found. From this clue, one can refer to the *Manual of Classification* and *Classification Definitions* to define a field of search more exactly.

A partial page reprinted from the *Index of Classification* is provided as an illustration. Index titles are arranged alphabetically, and the titles may be single words or phrases. When a title is a phrase, there are usually indented titles under the phrase which further define the subject matter.

The *Index* is arranged alphabetically with subheadings that can extend four levels of indentation. A complete reading of a subheading includes the title of the most adjacent higher heading outdented thereabove, the next most adjacent higher heading, and so on until there are no higher headings. An example is taken from the sample page:

"Azines, Azo compounds, Heavy metal containing."

An *Index* title is followed by a citation of a class and subclass within the classification system in which the subject matter relevant to the title is located. A plus sign (+) with the subclass citation indicates that relevant subject matter is located in that subclass and all indented subclasses thereunder.

The listings in the *Index* include unofficial subclasses, digests, and cross-reference art collections; these are designated by certain symbols and abbreviations set forth in the *Index*.

Manual of Classification

The *Manual of Classification* is a numerical listing of class titles with subclass titles under each class. Presently, there are about 305 classes and 73,000 subclasses, and the *Manual* in total consists of about 1,000 pages. It is a key publication in using the USPCS. The *Manual of Classification* contains the following:

1. A list of the current contents of the *Manual of Classification* showing the current page date for each class and the year in which the class was originally classified
2. A list showing an organization of science and technology into three main groups
3. A hierarchical arrangement of class titles in each main group by related subject matter
4. A list in numerical order of each examining group and art unit indicating examining group personnel, their location, and phone numbers
5. A list in numerical order by art unit indicating the classification assigned each
6. A list of classifications, in numerical order by class number, giving the class title, the art unit to which it is assigned, and the examiner search room in which it can be found
7. A list of classes in alphabetical order by class title giving the class number indicating where its schedule can be found in the *Manual*
8. The class schedule for Plants
9. Class schedules (listing of the subclass titles and numbers arranged in organized order for each class) arranged in the numerical sequence by class number
10. The design classes

The full title of a subclass in a given class schedule includes the following:

1. The class title
2. The title of most adjacent higher subclass in the first vertical column at the extreme left
3. The title of the next adjacent higher subclass indented in the second vertical column
4. The title of the next adjacent higher subclass indented in the third vertical column

As discussed in a previous section, utility is the primary basis of classification; built into the basis of classification is the concept of proximate function. The statutory categories of invention are product, process, apparatus, composition of matter, and certain varieties of plants. Classes are the systematic arrangement of subject matter on the basis of any one or more of these elements.

The subclasses in each class are arranged in order of complexity or comprehensiveness of the inventions included within the class. The complex wholes such as the "special" work and combined machines or processes come first, followed in order by simple whole machines, the major parts, the minor parts, or details and accessories, and the residuum or "miscellaneous."

Preceding subclasses are deemed "superior" to succeeding subclasses. The basic rule is to assign one patent copy (original) corresponding to a claimed invention to the first appearing subclass of a series of coordinate subclasses that may properly receive it and to place additional (cross-reference) copies where appropriate in the later appearing subclass or subclasses adapted to receive its parts.

In a class having the subclasses arranged according to this rule, no "original" patent copy should be found in any coordinate subclass following the one properly entitled to receive it, although it may be found either in preceding subclasses as a part of some other organization or in an indented more specific subclass.

A "rule" of superiority and inferiority applies also to patents containing subject matter classifiable in more than one class, the patent being assigned to the class designed to receive the largest claimed combination with a cross-reference, if necessary, for other claimed disclosures to the more elemental classes. Between classes, however, the positional relation (hierarchy) that exists between subclasses of the same class does not obtain, and "superiority" and "inferiority" is given only by the definitions.

Class and Subclass Definitions

The definitions are a definitive statement on the scope of the subject matter encompassed by a respective class or subclass. In addition, a majority of the definitions have accompanying notes. These notes are in general of two types: notes which supplement definitions by defining terms, giving examples, etc., and notes referring to a closely related disclosure in another class or subclass. The latter type of notes referring to related disclosure in other classes or subclasses are commonly called "search notes" and can be very beneficial. These notes, in effect, set the limits on a class or subclass in that these notes generally state the relationship to other classes and subclasses. The purpose of these notes is to guide a searcher to the extent necessary to reach a decision whether to include or exclude a class or subclasses in his field of search.

RETRIEVING INFORMATION

In preparing to search the collection of United States patents, it is first of all important to understand what you are searching for.

The first step of the search should be to prepare a "search summary" of the subject matter that you wish to search. It is suggested that one summarize the subject matter being searched in one or two paragraphs that are kept available during the search. This summary is useful in organizing the field of search and refreshing one's memory during the search.

Outlining a "field of search" will vary slightly in procedure depending upon whether one is familiar with the USPCS systems. If the searcher is not familiar with the general location of specific subject matter within the system, he should first scan the *Index to Classification*. From the alphabetical titles, one can usually identify classes or subclasses in the system that appear to be pertinent to a specific search problem. Noting the class and subclasses, the searcher proceeds to the steps outlined below.

One more familiar with the system, who has a general knowledge of the classes, may wish instead to scan the class titles which are listed in both alphabetical and numerical order in the front of the *Manual of Classification*. Noting the class and subclasses that appear to be pertinent to the search problem, he proceeds with the steps outlined below.

(1) Scan the class titles in the alphabetical or numerical list in the front of the *Manual* and note those classes that appear to include an answer to the search problem

(2) Study the class definition and the notes of the class which has a title that appears to answer the problem the best

(3) Scan the schedule of subclasses in the selected class and investigate first-line coordinate subclasses only

(4) Select for further investigation that first-line subclass which has a title and definition that best appear to include the solution to the search problem

(5) Carry the investigation to subclasses indented under the subclass selected in step 4

(6) Select the ultimately indented subclass which has a title and definition that best appear to include the answer to the search problem

(7) Inspect the prior art stored in the subclasses which appears the most likely to be productive and select those copies of the prior art that correspond to the search problem

(8) If the above subclass proves to be fruitless, continue the search in the parent subclass, that is, the subclass higher in the schedule

(9) If the above proves fruitless, repeat steps 4 through 8 with another first-line subclass selected by additional scanning of the class schedule

Failure to locate relevant prior art, although not necessarily anticipatory, usually indicates that the proper place in the system has not been reached. The examiners are available for consultation in outlining a field of search. In addition, the patent classifiers will assist in preparing a field of search when appropriate steps are taken.

At times it is necessary to conduct exhaustive searches as in the case of infringement searches, validity searches, etc. The approach is to search as outlined in the above paragraph, but also to extend the search upwardly above the proper locus of a specific search problem into areas in which the subject matter of the search might be found in combination with some other subject matter.

In many instances, cross-reference copies in a given search area provide the only clue to the location of subject matter of interest located elsewhere. In these cases, it may be very advantageous for a searcher to determine the original classification of a pertinent cross-reference for extension of a search into the segment of the system from which such cross-reference was made.

PATENT OFFICE DATABASE

The United States Patent Office classified search files currently consist of about 18 million United States patent documents, about 12 million foreign patent documents, and about 2 million documents of nonpatent literature. Each year an additional 75,000 new United States patents issue and are added to the file. With cross-referencing, a total of about 250,000 additional documents are added annually.

In addition, the Patent Office receives over 600,000 foreign patent documents each year. These documents are not classified by the USPCS when received and must be classified by the Patent Office. Of the 600,000 documents, approximately fifty percent are involved in "patent families" and are not included in the search files. An additional 50,000 nonpatent documents are classified each year for the search files.

Foreign patents and nonpatent documents are filed in the examiners' search files only, not in the classified files of the public search room. Actually, the examiners' search files are not completely filed with foreign patents and nonpatent literature. The examiners' search files contain only a percentage of the total foreign patent documentation that could by definition be included in these files. The Classification Division simply does not have adequate personnel or funds to classify completely all foreign patent documents. A disadvantage of the USPCS is that foreign countries do not classify their documents by the USPCS for distribution in this country. Therefore, one should never assume that since an anticipatory document does not exist in the examiners' search file, no such document exists. It could exist in the foreign patent collection or the nonpatent literature not included in these search files.

The Patent and Trademark Office has a computerized database called the

Automated Patent System which covers all U.S. patents from 1971 to date. This database is available at Patent Depository libraries throughout the U.S. as discussed in a later chapter. When the database is used in combination with the Image search system, it is possible to display the entire patent document, including the drawings.

Chapter 14

THE INTERNATIONAL PATENT CLASSIFICATION SYSTEM

INTRODUCTION

The International Patent Classification System (IPCS) is used by most major patent countries except the United States. Presently, about fifty countries classify published patent documents by this system. As discussed in the previous chapter, the United States classifies patent documents by the USPCS for its classified search files. Since 1969 the United States has in addition classified issuing patents by the IPCS for foreign dissemination; therefore, United States patents issued since 1969 contain both a United States classification and an international classification. Unfortunately, foreign countries do not reciprocate; they do not classify their patent documents by the USPCS.

The framework for the IPCS originated in 1954 in an international treaty known as the "European Convention on the International Classification of Patents for Invention" under the aegis of the Council of Europe. In 1967, the United International Bureau for the Protection of Intellectual Property, the predecessor of the World Intellectual Property Organization (WIPO), and the Council of Europe entered into negotiations for continuing work on the classification system. In 1971, a new treaty, the "Strasbourg Agreement Concerning the International Patent Classification," was negotiated by the WIPO and the Council of Europe. By this agreement, in 1975 the IPCS became a self-administering system no longer under the Council of Europe. Any country party to the Paris Convention for the Protection of Industrial Property may become party to the 1971 agreement.

The IPCS is under the responsibility of the World Intellectual Property Organization. By this structure, the IPCS is a worldwide system administered by an intergovernmental organization of worldwide scope. This chapter is concerned with the general aspects of this classification system.

COUNTRIES USING THE SYSTEM

The IPCS Strasbourg Agreement of 1971 requires that countries utilizing the system classify documents down to the finest subdivision when possible. At present, 36 countries and one international organization do this by classifying printed patent documents down to the finest subdivision of the system.

When the national procedure in a country does not search the "state of the art" when examining patent documents, classification to the subclass level is sufficient. At present, 11 countries classify their published patent documents down to the subclass of the IPCS system.

STRUCTURE

The IPCS is a hierarchical system consisting of six classification levels. These levels are listed below in hierarchical order along with the number of units within each level.

> Sections 8
> Subsections 29
> Classes 118
> Subclasses 617
> Main groups 7,000
> Subgroups 49,000 (approximate)

SECTIONS. The highest classification level is the section, which is a subdivision of the totality of technical knowledge. Each of the eight sections has a title and symbol; the title is a one or two word description, and the symbol is a capital letter of the alphabet. The eight sections are listed below:

> A Human Necessities
> B Performing Operations; Transporting
> C Chemistry and Metallurgy
> D Textiles and Paper
> E Fixed Constructions
> F Mechanical Engineering; Lighting; Heating;
>   Weapons; Blasting
> G Physics
> H Electricity

SUBSECTIONS. The subject matter in each section is divided into several subsections. The subsections consist of a descriptive heading for information purposes. The subsections have titles only, no symbol. The titles consist of one or more subscriptive words relating to the subsection. Usually there are only four or five subsections within a section. For example, section A, "Human Necessities," has the following four subsections:

Agriculture
Foodstuffs and Tobacco
Personal and Domestic Articles
Health and Amusement

CLASSES. The sections are subdivided into classes. Each class has a title and a symbol; the title consists of one or more descriptive words, and the symbol consists of the pertinent symbol of the section followed by a two-digit number. For example, the subsection "Foodstuffs and Tobacco" has the following four classes:

A 21 Baking; Edible dough
A 22 Butchering; Meat treatment; Processing poultry or fish
A 23 Foods or foodstuffs; Their treatment not included in other classes
A 24 Tobacco; Cigars; Cigarettes; Smokers' requisites

SUBCLASSES. Each subclass has a title and symbol; the title consists of a descriptive phrase, and the symbol consists of the symbol of the pertinent class followed by a capital letter of the Roman alphabet. For example, class A 21, "Baking; Edible dough," is divided into the following three subclasses:

A 21 B Bakers' ovens; Machines or equipment for baking
A 21 C Machines and equipment for making and processing dough; Handling baked articles made from dough
A 21 D Treatment, e.g., preservation, of flour or dough

GROUPS. The main subdivisions of the subclass are main groups. These are designated by a 1–3 digit number, an oblique stroke, and the number 00 in addition to the subclass symbol. For example, under the subclass A 21 B, "Bakers' ovens," is found

A 21 B 2/00 Baking apparatus employing high-frequency or infrared heating

SUBGROUPS. A majority of the main groups are further subdivided into subgroups. The numbering of a subgroup consists of a 1–3 digit number of the main group to which it constitutes a subdivision, the oblique stroke, and at least two digits other than 00 in addition to the subclass symbol. For example, the main group "Bakers' ovens" is divided into nineteen subgroups; the first four are listed below:

A 21 B 1/02 . Characterized by the heating arrangements
A 21 B 1/04 . . Ovens heated by fire before baking only
A 21 B 1/06 . . Ovens heated by radiators
A 21 B 1/08 . . . by steam-heated radiators

The title of each subgroup is preceded by one or more dots, and the number of dots determines the hierarchial level of the subgroup. For example:

1/00 T
1/02 . T1

1/04 . . T2
1/06 . T3

The general principle followed in the hierarchical sequence is that subject matter covered by a main group is divided into several one-dot groups; the subject matter covered by each one-dot group is divided into several two-dot subgroups, etc., up to seven dots. The scope of a subgroup is determined by the directly preceding superior subgroups. For example:

$$1/06 \; T \; + \; T \; + \; T3$$
$$1/10 \; T \; + \; T4 \; + \; Ts$$

The scope of subgroup 1/06 in the above example is determined by subgroups that precede it hierarchically, that is, the higher subgroups of 1/00 and 1/02. The scope of subgroup 1/10 is determined by subgroups 1/00 and 1/04.

To summarize the above discussion, a complete classification would consist of the following elements:

In general, the scope of a division is defined by its title. This system does not have the elaborate definitions occurring in the United States Patent Classification System. Occasionally, there are notes and or references associating a division with a second division, section, class, subclass, etc.

REFERENCES. In some cases, a class, subclass, or group title is followed by a phrase in brackets which refers the searcher to another place in IPCS.

NOTES. As stated above, the scope of a place in the IPCS is normally defined by its title. However, in some cases additional information is necessary to define exactly the scope of the location. This information is given in the form of notes associated with a section, subsection, class, subclass, or group.

PRINCIPLES

The IPCS uses the combination of two main classification principles: the first approach is the "function" approach as used by the United States system, where inventions are classified according to the function to which they are relevant; the second approach is the "industry" approach, where inventions are classified according to the branch of industry, art, or human activity to which they are relevant.

The IPCS is in principle mainly a function-oriented classification system. When possible, the function approach takes precedence over the industry approach. Take for example the activities of conveying, packing, storing, hoisting, lifting, and hauling. These are functions which deal with every branch of industry and would therefore be classified by the "function" approach since the function approach takes precedence over the industry approach.

On the other hand, some functions are exclusively relevant to certain branches of industry. For example, spinning, weaving, and knitting are concerned mainly with textiles. In this case, the industry-oriented approach would apply.

Some inventions disclose both intrinsic function and application. In these cases, IPCS rules allow multiple classification. In essence, it is up to the judgment of office personnel to choose in these cases between the two approaches or to choose a multiple classification.

RETRIEVING INFORMATION

The IPCS is designed to be used in different types of searches, not only novelty searches by the examiners in patent offices, but also searches by the public. Many people feel that the IPCS is an easier system for public use than is the USPCS.

Publications

There are four retrieval reference tools for retrieving technical information from the IPCS files. These include:

1. *Advice to Searchers*
2. *International Classification of Patents*
3. *The Guide to the International Patent Classification*
4. *Catchword Index*

Copies of these publications can be obtained in either French or English from Carl Heymans Verlog KG, Steinsdorfstrasse 10, Postfach 275, Munich, Germany.

ADVICE TO SEARCHERS. *Advice to Searchers* is a publication that provides detailed guidance on how to conduct a search in the IPCS. It also describes the various types of searches and how to prepare for such searches. This booklet should be read by everyone who needs to use the IPCS but is not familiar with its structure, content, and reference tools.

INTERNATIONAL CLASSIFICATION OF PATENTS. The *International*

Classification of Patents is the authentic text of the classification set forth in the Strasbourg Agreement concerning the IPCS. The revised version, presently in the fourth edition, was adopted in 1979 by the IPC Committee of Experts and entered into force on January 1, 1980. There are two versions, one in English and one in French.

The *International Classification of Patents* consists of nine volumes, 1,000 pages, available in bound or loose-leaf form. This is the key retrieval tool required for locating specific information in the IPCS. In addition to the collection of sections, classes, subclasses, groups, and subgroups previously discussed, this publication has a number of other features, including title references, notes, scope, subclass indexes, glossary, etc.

GUIDE TO THE INTERNATIONAL PATENT CLASSIFICATION. The *Guide to the International Patent Classification* is published as volume 9 of the *International Classification of Patents*. The purpose of the *Guide* is to explain the layout, use of symbols, principles, rules, and applications of the classification as they appear in volumes 1 to 8. This publication also gives advice on how to classify documents according to the International Patent Classification.

CATCHWORD INDEX. The *Catchword Index* is another essential reference tool for retrieving information from the IPCS. It comprises an extensive alphabetical listing of subject matter headings and descriptors. The purpose of the *Catchword Index* is to serve as an initial means of entry into the IPCS by providing the searcher with a specific citation of subclass group and subgroup relevant to the user's particular area of interest. Each catchword or subcatchword indicates the place in the International Patent Classification which deals with the subject in question.

Steps

Listed below are the steps one would follow in retrieving information from the IPCS.

1. Write a search summary of the subject matter of the search.
2. Consult the *Catchword Index*. Each catchword indicates the place in the IPCS which deals with the subject in question.
3. Read the appropriate place in the *International Classification of Patents*. The class and subclass titles, references, and notes elaborate on the subject matter in each. By reading these, one can confirm the main group selection.
4. Choose the most indented subgroup. After having selected a main group, scan all one-dot groups under the selected main group, identifying the ones most appropriate; repeat the step to choose the most appropriate two-dot subgroup, three-dot subgroup, etc.
5. Consider searching superordinate groups (fewer dots) under which the chosen group is located. This would widen subject matter coverage.

If one is unable to obtain a lead from the *Catchword Index*, he might try consulting the "contents of section" appearing at the beginning of each section in the *International Classification of Patents*. Scan the eight sections to select possible subsections and classes by title. After determining the appropriate main group in the subclass by reference to the subclass index, continue with step 4 above.

RECLASSIFICATION

A number of industrial property offices are now reclassifying according to the IPCS the patent documents published by them before the introduction of the IPCS in 1975. Some offices are also reclassifying patent documents kept in their collections for search purposes that are published by other industrial property offices but are not yet classified according to the IPCS. To make the results of this effort accessible to others, the WIPO signed in 1975 a document establishing the International Patent Documents Reclassified According to the IPCS (CAPRI) system. The aim of the CAPRI system is to collect and store the symbols of the IPCS allotted to patent documents issued before 1975. In view of the very high number of such documents, priority is being given to the coverage by the CAPRI system of those patent documents which constitute the "minimum documentation" under the Patent Cooperation Treaty. This includes patent documents issued from 1920 onwards by France, Germany, Japan, the Soviet Union, Switzerland, the United Kingdom, and the United States of America. The European Patent Office and the Japanese Patent Office are now cooperating in the CAPRI system.

Chapter 15

SEARCHING PATENT DOCUMENTS

INTRODUCTION

United States patents are classified by the Patent Office according to subject matter into about 305 classes and 73,000 subclasses. By using the *Manual of Classification*, one can identify classes and subclasses of subject matter of interest. After identifying the class and subclass of interest, the documents in these subdivisions can be retrieved for reading.

The United States Patent Office maintains a classified file of United States patents, arranged by subject matter, at the public search room in Arlington, Virginia. The searcher can remove the selected subclass bundles of patents from the open stacks, take them to a desk, and read the patent documents, which are arranged in chronological order. By using the classification system, one can retrieve a file of a few hundred patents in a specified subject area from the total base of five million plus documents.

Also at this location the Patent Office maintains a scientific library for public use. This library maintains a collection of scientific books, journals, and indexes to the scientific literature. In addition, the library contains journals of foreign patent offices with many foreign patent copies. Except for the examiners' search files, foreign patents are not filed by subject matter but are arranged in numerical order.

An additional classified patent file is maintained at various locations throughout the Patent Office complex for use by the patent examiners. In addition to the subclass patents, these files often contain unofficial subclasses, foreign patents, and some literature references classified by subject matter. Also, the examiners' search files sometimes have "digests" of subject matter. These materials can be very useful in searching, and with permission members of the public can use the examiners' search rooms.

The public search room is geographically inconvenient for regular use by most people. Certain libraries located throughout the United States are designated as "depository libraries" and maintain a United States patent collection and receive current issues of patents. These libraries are located in various states and are open for public use. Most of these locations have access to a computerized database of U.S. patents issued since 1971, PAT FILES, provided and maintained by the U.S. Patent Office.

To follow issuing patents, there are a number of abstracting services available. The *Gazette* is the official abstracting publication of the Patent Office; it appears weekly. The *Gazette* provides a sample claim and a drawing of all patents issuing that week. By following the *Gazette*, one can keep abreast with weekly issues in classes and subclasses of interest.

Over 20,000 patent documents are published each week. There are a number of abstracting services which prescreen, classify, and make available abstracts of these documents. Prominent among these are Chemical Abstracts, Derwent Patent Information, and Pergamon Information services. Other vendors such as STN and Dialog provide on-line access to their databases through their unique searching protocol. These databases cover U.S. patents, European patents issued under the Patent Cooperation Treaty, and other patents from around the world collected under the auspices of the World Intellectual Property Organization. The services of these and other similar companies are discussed in this chapter.

AT THE PATENT OFFICE

The Patent and Trademark Office is located at Crystal City, Arlington, Virginia, where it maintains a public search room and scientific library for public use. The public search room is a library-type facility with copies of United States patent documents. There is ample unreserved seating for searching patents and taking notes, and token-operated photographic equipment is available for making copies.

The classified files of United States patents are filed in boxes called "shoes" adjacent to the public search room. The classified files contain all issued United States patents (approximately 5,200,000) arranged according to subject matter. In addition, approximately 8,000,000 copies are cross-referenced to related technical fields in the files.

On the second floor there is a numerical file of United States patents. Utility patents are numerically arranged on microfilm in a series of labeled cartridges. Patents since 1,999,760 are also available in bound volumes. All reissue patents and design patents numbered over 10,000 are on microfilm.

There is also a numerical list of United States patents and their classification which gives both the original classification and all cross-reference classifications for each patent. This is available on microfilm and a computer display board.

Public Search Room

Every person using the public search room, scientific library, or examiners search room at the Patent Office must obtain a user's pass from the clerk in the public search room. In accepting the user pass, the searcher agrees to follow all regulations established by the commissioner of patents for using these facilities.

The stepwise procedure for determining a "field of search" was discussed in Chapter 13. Copies of the three publications used in determining the field of search, the *Index to Classification, Manual of Classification,* and *Definitions of the Classes and Subclasses,* are found on tables as one enters the public search room.

The search room has a staff of search advisers trained to answer questions on search problems. These advisers are located in the lobby of the search room. Obviously these advisers are not experts in every subject area, but they can usually indicate a starting point for a search. The advisers are not allowed to perform any part of the search, but they will answer questions.

After establishing a "preliminary field of search," one has a last valuable resource to consult before the actual search. The *Manual of Patent Examining Procedures,* Sect. 713.02, permits examiners to indicate a field of search to a searcher. Before beginning the search, one should verify the field of search with an examiner. The examiner will either confirm that the field of search is correct, or he may suggest another field or another examiner to consult regarding the field of search.

The appropriate examiner is found by using the *Manual of Classification.* The *Manual of Classification* contains a list of classes arranged numerically associating each class with the group art unit in charge of that class. From the preliminary field of search, one can identify the art unit associated with the subject matter of the invention. In addition, there is an index wheel outside the public search room in the hall relating each class and subclass to the appropriate group art unit. The location of the examining groups is posted in the lobbies of buildings 3 and 4. The clerk of each group will advise the searcher of the examiner in charge of a particular subclass.

The examiner's participation in establishing a field of search is a courtesy, not a matter of right. One should remember this when approaching the examiners; courtesy will get the searcher a lot of help, a lack of courtesy will get him thrown out of the examiner's office. A searcher should also have a written preliminary field of search when approaching an examiner, never approach an examiner unprepared.

After verifying the field of search with the appropriate examiner, the searcher returns to the public search room to begin the search. This is done by securing a reasonable number of bundles of patents in the subclass of the established field of search for reading. The search advisers will advise the searcher of the location of any particular subclass. Also there is a turn wheel just outside the search advisers' office containing this information. There are carts available for transporting the boxes of patents from the stacks to the reading tables.

Unfortunately, patent copies are sometimes removed from the shoes by certain persons. If the search is of particular importance, the searcher may wish to obtain a list of patents assigned to each subclass. This information is available on microfilm on the second floor, and copies can be made. By this method, the searcher can verify that each subclass searched was complete at the time of searching.

A disadvantage of searching at the public search room is that often an entire class or subclass is removed from the stacks by the classification division for reclassification. It is not uncommon to have as high as five percent of the total collection removed at any time for reclassification.

Computer Databases

The Patent Office has a database of all U.S. patents (approximately 1,800,000) issued since 1971; it is called the Automated Patent System (APS). When using the Image Search system, the APS has the ability to display the full text and drawings and other illustrative material. The system was designed primarily for the examiners to use in patentability searching. The origin of the database is the magnetic tapes which the Patent Office started to use in 1971 to print patents; prior to that time patent documents were typeset.

This database is made available to certain depository libraries throughout the United States, as discussed in a following section. Also, the Patent Office allows certain professional patent search firms to access the database. These firms provide patent searches for the public at a fee.

The Patent Office maintains several other databases which can be used by the public at the search room, and they are provided to the patent depository libraries. These are discussed in a following section on depository libraries.

Examiners' Search Room

An additional classified patent file is maintained at various locations throughout the Patent Office complex for use by the Patent and Trademark Examining Corps. With permission, members of the public can search the examiners' search files between 8:45 A.M. and 4:45 P.M. on work days.

The examiners' search files contain additional documents, including unofficial cross-references, foreign patents, and nonpatent documents arranged by class and subclass. In some files the subclasses are divided into unofficial subclasses as discussed in Chapter 13. In addition, some classes have patent collections in addition to subclasses. These collections are called "digests," and are most often filed at the end of the subclass.

Searching in the examiners' search files offers the advantage of additional documents classified by subject matter, but, there is a disadvantage in that documents are often removed by the examiner from the files of cases under examination. One is therefore never certain that the boxes of patent copies being searched are complete. One solution is to check a list of United States patents assigned as original or cross-referenced to the subclass. This information is available on microfilm and can be copied.

To use the examiners' search files, determine the location of the examining group having the class or subclass of interest by referring to the flip chart at the search room directories. Thereafter go to the receptionist for the examining group and sign in.

The "shoe case" cabinets for patent copies are labeled for each class and subclass. Usually foreign patents and nonpatent documents are at the end of the subclass file.

While these files do contain certain foreign patent documents, they are not complete. Non-English language patent documents and non–English language publications are fairly uncommon in the examiners' search files. Therefore, one should never assume that since a document is not present in the examiners' search files, no such document exists. If the search is of particular importance, one may wish to have a foreign search conducted as discussed later in the chapter.

Scientific Library

The scientific library is located on the second floor above the public search room. The scientific library contains a collection of scientific books, journals, and foreign patent files for all major foreign countries. While the scientific library has a good collection of indexes to the scientific literature, the collection of scientific literature itself is quite modest.

Copies of foreign patents are arranged in numerical order. Some copies are bound volumes, and others are on microfilm. Token-operated photocopy equipment is available for both paper and microfilm. A reading room is available for use of these materials.

Generally, the scientific library is not the best place to search either scientific literature or foreign patents. The literature collection is quite limited, and one would probably be better served by a university library. While many foreign patent documents are available, they are not filed by subject matter and thus are of little value unless one is looking for a particular document from a known country with a known patent number.

PATENT DEPOSITORY LIBRARIES

Certain libraries throughout the country are designated as patent depository libraries and receive current issues of all United States patents and maintain a collection of earlier issued patents. The extent of the collection varies from library to library; some collections are for recent issues, while others have documents back to 1870. With one exception (Sunnyvale Patent Library), these collections are by patent number sequence, and not by subject matter as at the public search room.

These patent collections are open for public use and have facilities for making copies. Depending on the library, the copies may be bound volumes, on microfilm, or a combination of both. The libraries usually have copies of other Patent Office publications such as the *Manual of Classification, Index to the United States Patent Classification,* and *Definitions,* and they provide staff assistance in the use of these.

In an effort to make patent and trademark information more readily available to inventors and entrepreneurs throughout the United States, the Patent Office maintains several basic databases which are usable at the patent depository libraries free of charge. One of these databases is on-line from the Patent Office; others are distributed on compact disc–read only memory (CD-ROM). At present there are 17 major files on 27 different CD-ROM discs. The author would like to thank Jean Potter, head of the documents department, North Carolina State University libraries, for her assistance with the following information.

Databases on CD-ROM

There are six different databases on CD-ROM, including CASSIA, ASSIST, SNAP, TRADEMARKS, JAPANESE PATENT ABSTRACTS, and IMAGE. Search results from these may be printed out or down-loaded to diskette.

The CASSIA — Classification and Search Support Information System — contains three files. The CASSIA/BIB disc contains titles, classification, dates, state or country of origin, and assignee codes for patents issued from 1969 to present. There are approximately two and one half years worth of abstracts on the CD-ROM. Searches may be done by assignee code, classifications, and or words in the titles and abstracts. The CASSIA/CLSF disc contains patent numbers and classification for all U.S. patents issued since 1790. The CASSIA/ASIGN disc contains assignment transactions since 1969.

The ASIST disc contains several independent files. The discs are updated about four times per year and contain the following:

1. Patentee-Assignee File, which includes assignees for utility patents from 1968 and other types of patents from 1977, as well as patentees from 1975

2. Attorney/Agent Roster File, which is a listing of persons registered to practice before the Patent Office

3. Index to the U.S. Patent Classification, an alphabetical index used in combination with the *Manual of Classification*

4. *Manual of Classification*, which is a hierarchical arrangement of all classifications assigned by the Patent Office

5. *Manual of Patent Examining Procedure*, which is a procedural guide for the examiners

6. *Classification Definitions*, which are definitions for the 300 plus classes and subclasses

7. *Classification Orders Index*, which identifies order numbers related to classification changes since 1976

8. *USPCS to IPCS Concordance*, which correlates the U.S. patent classification scheme to the international classification scheme

9. Art Unit Search Room Locator, which identifies the location of shoes for particular classifications at the public search room

The SNAP — Serial Numbers for Allowed Patents — is a file which allows the retrieval of issued patents by application serial numbers for applications filed since 1976.

The TRADEMARK file lists active and pending trademarks and assignees of trademarks.

The JAPANESE PATENT ABSTRACTS file includes eight discs with English language abstracts of published unexamined patents (KOKAI) for the years 1980–82 and 1984–90.

The IMAGE discs sets contain full-text images of patents created for the subject matter of acid rain and genetic engineering.

On-Line Databases

Since September of 1991, the Patent Office has provided on-line access to the text of its Automated Patent System to a limited number of the depository libraries. The Automated Patent System (APS) is a full-text database containing all U.S. patents from 1971 to present. The system was developed for in-house examiner use, but access is allowed to certain subscribers. At present fourteen depository libraries are on line. During the fiscal year 1994-95, the APS will be offered to depository libraries.

The APS database is searchable by keywords, inventor, assignee, various dates, references cited, and more. The database is updated nightly. No drawings are available on the APS, text only. This is unlike the Patent Office's PAT FILE, which can be combined with the IMAGE database for drawings.

For the first few years, the Patent Office has offered access free of charge. However, as other libraries come on line, fees will be charged.

Listed below are the locations of the current patent depository libraries.

STATE	NAME OF LIBRARY
Alabama	Birmingham Public Library
California	Los Angeles Public Library Sacramento: California State Library Sunnyvale Patent Library
Colorado	Denver Public Library
Georgia	Atlanta: Price Gilbert Memorial Library, Georgia Institute of Technology
Illinois	Chicago Public Library
Massachusetts	Boston Public Library
Michigan	Detroit Public Library
Missouri	Kansas City: Linda Hall Library St. Louis Public Library
Nebraska	Lincoln: University of Nebraska-Lincoln, Love Library
New Jersey	Newark Public Library
New York	Albany: New York State Library Buffalo and Erie County Public Library New York Public Library (The Research Libraries)
North Carolina	Raleigh: D.H. Hill Library, N.C. State University
Ohio	Cincinnati & Hamilton County Public Library Cleveland Public Library Columbus: Ohio State University Libraries Toledo/Lucas County Public Library
Oklahoma	Stillwater: Oklahoma State University Library
Pennsylvania	Philadelphia: Franklin Institute Library Pittsburgh: Carnegie Library of Pittsburgh University Park: The Pennsylvania State Libraries
Rhode Island	Providence Public Library
Tennessee	Memphis & Shelby County Public Library and Information Center
Texas	Dallas Public Library Houston: The Fondren Library, Rice University
Washington	Seattle: Engineering Library, University of Washington
Wisconsin	Madison: Kurt F. Wendt Engineering Library, University of Wisconsin Milwaukee Public Library

OFFICIAL GAZETTE

The *Official Gazette* is the official abstracting journal of the Patent Office. It is published weekly, issuing on Tuesday, and provides a selected claim and drawing of each patent issuing that week. By using the *Gazette*, one can keep up-to-date with issuing patents.

Patents listed in the *Gazette* are arranged in three major categories: general and mechanical, chemical, and electrical. In each category the patent claims are grouped in numerical sequence according to the United States Patent Classification System. Therefore to use the *Gazette* effectively, the user must identify the classes and subclasses of interest. Once this has been done, one can readily follow the patents issuing each week in those classes and subclasses. The *Gazette* can be found in most university and public libraries as well as through most industrial research organizations. It can also be purchased by subscription and single copy from the superintendent of documents.

Since the *Gazette* only lists the main claim and a drawing, it does not provide a lot of information. The major benefit of this information is that it helps one decide whether to obtain a full copy of the patent or at least to locate an abstract that would provide more information.

In addition, the *Gazette* contains information on other patent matters such as notices of patent and trademark suits, a listing of patents for sale or license, information relating to Patent Office rule changes, notices of certificates of correction and reissue applications, and an abstract and drawing from defensive publications.

The *Gazette* contains a weekly index, and these weekly indexes are combined to form the *Annual Index of Patents*. The *Annual Index* is in two volumes, one an index of patentees and the other an index by subject matter of patents. The subject index follows the patent classification system; the *Annual Index* does not contain a general subject index.

When one is searching for an inventor or organization assignee, the indexes make this search relatively easy. When looking for organizations, however, one must consider name changes because companies are often changing their names by mergers, subsidiaries, etc. The *Annual Index of Patents* is available in many libraries and is obtainable from the superintendent of documents. A disadvantage of this publication is that it is released many months after the end of the calendar year.

ABSTRACTING SERVICES

Over 20,000 patent documents are published weekly. There are several companies which prescreen, classify, and publish abstracts of patent documents.

Prominent among these are Chemical Abstracts, Derwent Publications, Pergamon International, and International Patent Documentation Center.

Chemical Abstracts

Chemical Abstracts (CA) is a major source of chemical and chemical engineering patent information. Chemical Abstracts indexes patent documents from 26 nations, including the European Patent Office and the World Intellectual Property Organization under the Patent Cooperation Treaty. Approximately 20% of the abstracts in *CA* relate to patent documents. *Chemical Abstracts* covers all patent documents of chemical or chemical engineering interest issued by the major patent countries, including the United States, the European Patent Office countries, and Japan. For the smaller patent countries, *CA* covers only chemical and chemical engineering patent documents issued to individuals or organizations of that country.

When patent rights for the same invention are applied for in more than one country, *CA* abstracts only the first patent document that it receives. Subsequent patent documents relating to the same invention are linked to the original abstract and other patents on the same invention through its *Patent Concordance*.

Indexing and abstracting of patent documents by *CA* are very good. The abstracts are usually concise, informative statements on the new disclosure. *Chemical Abstracts* tends to place emphasis on the technical disclosure rather than the legal aspects of the patent. There is usually about a year lag from the patent document publication date to an available abstract. Since *CA* is widely available in libraries and organizations, it is a major source for patent information. Also *CA* uses its own index systems rather than patent classifications systems and can thus be readily used by persons not familiar with the classification systems.

Derwent Patent Information Services

Derwent Publications, Ltd., a subsidiary of the Thomson Organization, is located at Rochdale House, 128 Theobalds Road, London, WClX8RP, England. It has an American office in Arlington, Virginia. Derwent abstracting services include:

1. *Derwent Central Patents Index* (CPI), which is an abstracting and retrieval service dealing with chemical and chemically related patent documents

2. *World Patents Index* (WPI), which is a collection of indexes from 24 countries giving details of all inventions classified according to subject matter, patentee, patent family, and patent number

3. *World Patents Abstracts* (WPA), which is a collection of abstracts of the inventions reported one week earlier in the *World Patents Index*

The *Central Patents Index* is an abstracting and retrieval service dealing with chemical-type patents under twelve subject matter categories. Basically, the *CPI* takes the form of three publications:

1. *Altering Abstracts,* which gives 100–200 word abstracts of current chemical inventions arranged by country
2. *Altering Classified Abstracts,* which appears a week later, contains the same abstracts arranged by the Derwent Classification System
3. *Basic Abstracts,* which appears one week later, gives a detailed, coded abstract of all first disclosures or basic patents

The database of the Derwent services includes twelve "major" and eleven "minor" countries. Among the countries included are the United States, the countries of the European Patent Office, and Japan.

As previously stated, the first *CPI* booklet is the *Alerting Abstract,* which is a 100–200 word abstract in English of current inventions arranged by country. The abstracts are quite informative. The description includes the local filing date, the application number, and the number of pages in the document abstracted.

There are twelve different editions of *Alerting Bulletins,* one for each of the twelve major countries. These booklets are published one week after the corresponding *World Patents Index Gazette,* which is published about six weeks after the actual issue date of the document abstracted.

A week after the country-sequenced *Alerting Bulletins* appear, the same abstracts are published arranged according to the Derwent subject classification system in the form of the *Classified Bulletins.* These bulletins are more suited to scientists who wish to be alerted to new inventions in their specialized field. By this format, one can follow only the invention disclosures in a specified field.

The *Basic Abstracts* are multicountry abstracts published one to two weeks after the *Alerting Bulletins;* they contain more detailed summaries of the invention than do the bulletins. There are twelve separate journals published weekly by Derwent, and these are published according to subject matter grouped by the twelve sections of the *CPI* classification system. The sections of the *CPI* classification include Plasdoc, Farmdoc/Agdoc, Food Fermentation, Disinfectants and Detergents, Chemdoc, Textiles, Paper and Cellulose, Printing, Coding and Photographic, Petroleum, Chemical Engineering, Nucleonics, Explosives and Protection, Glass, Ceramics, Electro(in)organic and Metallurgy, Catalysts.

The abstract is a summary of about 300 words and often includes examples and/or drawings. Typically, the abstract is broken down into paragraphs. The first paragraph usually defines the scope of the invention, the second paragraph usually makes a statement on use of the invention, and details of the invention follow.

The *World Patents Index* is a collection of indexes giving certain information for all inventions covered in the twenty-four countries. The information in the *World Patents Index* is classified according to patentee, subject matter, patent number, patent family, and priorities claimed. There are no abstracts included in the *WPI*.

The *WPI* Gazettes are published in the four editions: general, mechanical, electrical, and chemical. Each edition is based on the International Classification System and contains four indexes: patentee index, IPC index, accession number index, and patent number index.

The *WPI* contains the first announcement of a patent in any of the Derwent services, and this usually occurs four to six weeks after the actual issue date of the document reported. The *WPI* is useful for

1. Following subject matter by the International Classification system
2. Following patents by patentee
3. Determining if a given invention is being filed in another country

The *World Patents Abstracts* is a collection of abstracts of the inventions reported one week earlier in *WPI*. Currently, *WPA* offers two types of weekly publications: twelve separate publications related to particular countries and six publications dealing with different areas of technology on a multicountry basis.

For nonchemical subject matter, Derwent provides weekly coverage giving abstracts by country arranged by subject. Currently, there are seven different weekly editions, and these editions are for the most part based on the International Patent Classification System.

Each journal contains bound together in the weekly booklet all relevant pages from *Belgian Patents Abstracts, British Patents Abstracts, German Patents Abstracts, German Patents Gazette, Russian Inventions Illustrated,* and *United States Patents Abstracts,* and pertinent abstracts of European and PCT patent applications. Abstracts of basic patents in nonchemical fields are also available on microfilm.

Derwent's *World Patents Index* and several other databases are available on-line. Information can be obtained from the Derwent office.

Pergamon International Information Corporation

Pergamon International Information Corporation has offered for years an on-line information database and other related patent information products. The Pergamon patent database is available for direct computer terminal access. This service provides for on-line computer searching of abstracts, text, and certain bibliographic data from United States patents issued since 1970.

Just before this book went to press, the company was sold to the QUEST company. It is anticipated that similar services will be offered by QUEST. The database is United States patents issued since 1971. In 1971, the United States Patent Office started printing patent specifications from magnetic tape. The Patent Office has since offered these tapes to commercial companies for a fee.

International Patent Documentation Center

Patent applications are often filed in more than one country to form so-called patent families. In certain situations, it is desirable to determine all patent documents in a patent family. Most people believe the most complete patent equivalent coverage is offered by the International Patent Documentation Center (INPADOC) located in Vienna, Austria. In addition to the patent family service, INPADOC offers an international patent alerting service and patent inventor service.

The INPADOC database covers 48 countries, including all of the major patent offices. INPADOC services are offered by subscription and include references to all documents received by INPADOC. The time lag between issuance and publication is two to four weeks. The information is arranged in five sections: patent number by country, international patent class, assignee or company name, patent family, and patent inventor. INPADOC offers the following services:

PATENT FAMILY SERVICE. The *Patent Family Service* is a patent priority listing used to find equivalent documents. The *Patent Family Service* can be sorted by priority country, priority date, and priority number. Documents with more than one priority occur as many times as necessary. The *Patent Family Service* is issued monthly with cumulative updates.

PATENT INVENTOR SERVICE. The *Patent Inventor Service* can be used to obtain a complete microfiche listing of patents held by the inventor for the period 1968 to date. The user can determine areas of technology covered in terms of International Patent Classification. The *Patent Inventor Service* can be sorted alphabetically by inventor's name, IPC in ascending order, and country within IPC. Documents with more than one inventor's name occur as many times as necessary. The *Patent Inventor Service* is issued quarterly with cumulative updates.

INTERNATIONAL PATENT ALERTING SERVICE, *Patent Number Section*. The *Patent Number Section* of the International Patent Alerting Service is a numerical document list used to find documents of interest without additional information, e.g., publication date. The *Patent Number Section* can be sorted by publication country, document number, and publication date. The *Patent Number Section* issues weekly.

INTERNATIONAL PATENT ALERTING SERVICE, *International Patent Class Section*. The *IPC Section* of the International Patent Alerting Service may be used to keep abreast of changing technology. The user can set up IPC profiles and current awareness programs. The *IPC Section* can be sorted by IPC symbol, publication country, and publication date. The *International Patent Class Section* issues weekly.

INTERNATIONAL PATENT ALERTING SERVICE, *Assignee Section*. The *Assignee Section* of the International Patent Alerting Service may be used to keep current on companies of interest. The user can set up internal collections of assignees and update these files each week. The *Assignee Section* can be sorted by assignee name, main IPC symbol, and publication country. The *Assignee Section* issues weekly.

EXAMPLE OF A FIELD OF SEARCH

The process for determining a "field of search" was previously introduced in Chapter 13. Sections in that chapter discussed the three publications involved, *Index to Classification, Manual of Classification,* and *Classification Definitions,* and set forth steps for using these publications as outlined in DUPACS. This section provides an example on how an actual field of search might be delineated in the case of a chemical search problem; a similar approach would be used regardless of the subject matter.

STEP I. The first step is to determine the essential function or effect of the art or instrument of the search. One should prepare a "search summary" of the subject matter in the search in one or two paragraphs. This summary is useful in determining the field of search by refreshing the memory during the search.

Below is a search summary involving the industrial preparation of ethyl alcohol for which a field of search will be determined.

SEARCH SUMMARY. A method for making industrial ethanol (ethyl alcohol) is by hydration of ethylene under pressure at approximately 300°C with phosphoric acid catalysis according to the following equation:

$$H_3PO_4$$
$$CH_3CH_3 \qquad CH_3CH_2OH$$
$$300°C$$

By this procedure, a dilute solution of ethanol is obtained; distillation of this mixture results in an azeotropic mixture of 95.6% ethanol. To obtain pure ethanol, the 95.6% azeotropic mixture can be distilled in the presence of benzene, resulting in the separation of the water to give absolute ethanol.

In this example, let us assume that the invention concerns the process for removing the water from the 95.6% ethanol azeotropic mixture; the hydration of ethylene at elevated temperatures with phosphoric acid exists in the art.

Considering this, keywords in this search problem would include distillation, separation, and azeotropic.

STEP 2. Scan the major headings in the *Index to Classification* to obtain clues to pertinent classes and subclasses that include the subject matter being searched. An appropriate starting point would be "distillation" because this is a keyword in this search problem. Reproduced on page 172 is the appropriate segment taken from the *Index to Classification*.

Scanning the subheading titles under "Distillation," two subheadings, "Azeotropic" and "Separatory," appear to be most appropriate in this search. Both subheadings are associated with Class 203. Therefore, Class 203 DISTILLATION, PROCESS, SEPARATORY is taken as a starting point.

An alternate approach that one might use in step 2 is to scan the class titles in the alphabetical or numerical list in the front of the *Manual of Classification*. By this method, one notes the classes that appear to be relevant to the subject matter of the search problem.

STEP 3. Note the class titles that most specifically include the subject matter of the search. In this case, Class 203 DISTILLATION, PROCESS, SEPARATORY is taken as a starting point.

STEP 4. Inspect carefully the class definitions and notes of the selected class. As discussed in Chapter 13, class definitions set forth the metes and bounds of the subject matter of the class. In addition, class definitions sometimes contain possible exceptions to the principles of superiority of the subject matter in the class. When any deviations from the principles of superiority are mentioned, these are explained in the class notes.

Reproduced on page 173 pages from the *Classification Definitions*, Class 203, beginning with the mainline subclass 12, PROCESS REMOVING ONLY WATER FROM FEED MIXTURE.

Referring to the definitions and notes of mainline subclass 12, these suggest that the proper subclass is probably a subclass indented under this mainline subclass. In particular, note (2) states: "A plural distillation process of separating only water by adding an extraneous liquid to the distilland to alter the relative volatility of water and the liquid be dried in the initial distillation step. . . ."

STEP 5. If the class definitions indicate the selected class as the proper one, scan the schedule of subclasses of the selected class in the *Manual of Classification*. First inspect the major coordinate subclasses whose titles begin farthest to the left and have their initial letters arranged in vertical alignment.

Reproduced on page 174 is the schedule of subclasses for Class 203, DISTILLATION, PROCESS, SEPARATORY, taken from the *Manual of Classification*.

STEP 6. Select the first appearing major subclass title that includes a characteristic of the invention and study the definitions of the subclass.

As previously discussed, it appears that subclass 12, DISTILLING TO

Pertinent page from *Index to Classification*.

CLASSIFICATION DEFINITIONS

7. Processes under subclass 6 directed to adding a substance to inhibit or prevent corrosion of the apparatus and/or to inhibit or prevent scale formation.

SEARCH CLASS:
196. Mineral Oils: Apparatus, subclass 133, for a mineral oil vaporizer having some special feature of construction.
202. Distillation: Apparatus, subclass 267, for apparatus in terms of the materials of construction.
208. Mineral Oils: Processes and Products, subclass 47, for a process of treating mineral oil including a step to prevent or reduce corrosion or erosion of the apparatus employed in the process and see "Search This Class, Subclass" and "Search Class" thereunder for related fields of search.
252. Compositions, subclass 175+, for water-softening or purifying or scale-inhibiting agents, and subclass 387+, for anti-corrosion agents.

SEARCH CLASS:
55. Gas Separation, subclass 29+, for sorptively removing water from a gas, subclass 36+, for degasifying liquids and subclass 39 for a process for deaerating boiler feed water.
159. Concentrating Evaporators, subclass 5+, for concentrating apparatus of the film type, subclass 13+, for evaporating apparatus designed to maintain the liquid being evaporated in a film and subclass 49, for an evaporating process in which the liquid to be concentrated is spread in a thin film.
165. Heat Exchange, subclass 3, for a process of adding water vapor to air or removing water vapor from air.
202. Distillation: Apparatus, subclass 167, for a separatory distillation apparatus which includes a still and a feed-water heater.
210. Liquid Purification or Separation, subclass 1+, for a process of purifying a liquid not otherwise provided for.
423. Chemistry, Inorganic, subclass 580 for processes of producing water, e. g. heavy water including a chemical reaction.

8. Processes under subclass 6 directed to adding a substance to inhibit or prevent unwanted polymerization.

SEARCH THIS CLASS, SUBCLASS:
30. for a separatory distillation process in which a substance is added to cause a desired polymerization of at least one component.

11. Processes under subclass 10 in which distillation is carried out under a pressure greater than atmospheric or under a vacuum.

SEARCH THIS CLASS, SUBCLASS:
73+, for a plural distillation process in which at least one distillation is under pressure or vacuum and subclass 91+, for a single distillation process carried out under pressure or vacuum.

9. Processes under subclass 8 directed to inhibiting or preventing the polymerization of an unsaturated hydrocarbon.

12. Processes under the class definition in which only water is removed from the feed mixture. ←

10. Processes under the class definition of purifying water in which the only material recovered as a product is water.

(1) Note: For purposes of this and indented subclasses water is the impurity of the distilland which is to be removed. A process of removing other impurities as well as water is excluded.

Pages from *Classification Definitions*, Class 203.

CLASSIFICATION DEFINITIONS

(2) Note: A plural distillation process of separating only water by adding an extraneous liquid to the distilland to alter the relative volatility of water and the liquid be dried in the initial distillation step and then distilling a product of the initial distilling operation to separate the extraneous liquid is classified here.

SEARCH THIS CLASS, SUBCLASS:
　　10+, for a process in which the only material recovered as a product is water and 50+, for a process directed to adding a specific extraneous material to alter the relative volatility of a component of a mixture. ··

SEARCH CLASS:
　　23. Chemistry, subclass 306, for a process for concentrating a solution of a liquid in a liquid not otherwise provided for and subclass 274+, for apparatus for concentrating a liquid in a liquid.
　　34. Drying and Gas or Vapor Contact With Solids, appropriate subclasses, under "Processes" for a process for separating a liquid from a solid.

　　55, Gas Separation, subclass 20+ for a process of removing water from a gaseous or vaporous fluid.
　　62, Refrigeration, subclass 93+ for a process of removing moisture from air.

13.　Processes under subclass 12 in which the liquid substance is aqueous nitric acid.

SEARCH CLASS:
　　423, Chemistry, Inorganic, subclass 390+ for producing nitric acid by a chemical reaction.

14.　Processes under subclass 12 for separating water from an organic compound.

(1) Note: Mixtures of organic substances from which only water is separated by a distillation step are included in this and indented subclasses unless otherwise provided for.

SEARCH CLASS:
　　208. Mineral Oils: Processes and Products, subclass 187+, for a process for removing water from mineral oils and see "Note", "Search This Class, Subclass" and "Search Class" in subclass 187 for related processes for removing water from organic mixtures.
　　260. Chemistry, Carbon Compounds, appropriate subclasses, for a process including removing water by distillation combined with a step for forming a compound or extracting the compound from a natural source.

15.　Processes under subclass 14 in which the organic substance is an organic acid.

(1) Note: The term "organic acid" includes organic compounds which contain an acid function, e.g., boro, phospho, sulfo or carboxylic group and see Class 260, Chemistry, Carbon Compounds, subclass 500 "(1) Note".

SEARCH THIS CLASS, SUBCLASS:
　　14, for a separatory distillation process for removing only water from salts or esters of organic acids.

16.　Processes under subclass 15 in which the organic acid is acetic acid.

17.　Processes under subclass 14 in which the organic substance is an aldehyde or a ketone.

(1) Note: The terms "aldehyde" and "ketone" include those compounds having the structure R_1COR_2 wherein R_1 is hydrocarbon and R_2 is either hydrogen or hydrocarbon. See Class 260, Chemistry, Carbon Compounds, subclasses 586 and 598 and the notes thereunder.

18.   Processes under subclass 14 in which the organic substance is an alcohol.

(1) Note: For purposes of this and indented subclasses the term "alcohol" is limited to a hydroxy group bonded to carbon.

19.   Processes under subclass 18 in which the alcohol is ethanol.

20.   Processes under the class definition directed to defoaming or inhibiting the formation of foam.

SEARCH CLASS:
   55. Gas Separation. subclass 36. for a process of removing gas from a liquid and see (1) Note. "Search This Class, Subclass" and "Search Class" thereunder for related fields of search. subclass 87. for a gas separation process including the step of breaking foam and subclass 178. for apparatus in which the mixture is a substantially stable aggregation of gas or vapor bubbles dispersed in a liquid phase (foam) and comprising means to destroy or remove the aggregation and see "Search This Class, Subclass." and "Search Class" thereunder for related field of search.
   137. Fluid Handling, subclass 107.1+, for apparatus for controlling the degree of foaming in a gas charged liquid.
   195. Chemistry. Fermentation, subclass 107. for a fermentation process including the step of treating the foam produced.
   201. Distillation: Processes, Thermolytic, subclass 9. for a process of surface treating the solid particles of the charge to inhibit. reduce or prevent foaming during distillation.
   202. Distillation: Apparatus, subclass 264. for distillation apparatus intended to break foam. or inhibit foaming.

252. Compositions. subclass 321 for a process not combined with distillation for inhibiting foam and Search Class thereunder; subclass 358. for compositions for use in breaking colloids; subclass 361. for apparatus means for breaking foam.

21.   Processes under the class definition directed to recovering waste heat by indirect heat exchange with (1) a disparate source or (2) a product of a distillation step.

(1) Note: Heat generated by an engine which runs a compressor used in the process is a disparate source of "waste heat" within the scope of this and indented subclasses.

SEARCH THIS CLASS. SUBCLASS:
   100. for a digest of distillation processes directed to specific types of heating.

SEARCH CLASS:
   34. Drying and Gas or Vapor Contact With Solids. subclasses 19 and 35. for a process including conserving heat by indirect heat exchange.
   55. Gas Separation. subclass 80+, for a process of gas separation including indirect heat transfer and under subclass 80 see "Search This Class. Subclass" and "Search Class".
   62. Refrigeration. subclass 96, for a process in which heat from a gas being cooled is transferred to a heat absorber by indirect heat exchange, and subclass 113. for a process of refrigeration in which one function is in heat exchange relation with a second function.
   165. Heat Exchange. appropriate subclasses. for heat exchange apparatus and note Search Class under Class Definition for related fields of search.
   196. Mineral Oils: Apparatus, subclass 134. for apparatus for vaporizing mineral oils including means for heat recovery from the vapor or residuum.

SEPARATE OR REMOVE ONLY WATER, is the most pertinent. Indeed, the definitions and notes suggest this.

STEP 7. If the major subclass definitions include the subject matter of the invention, scan the schedule of minor subclasses indented one place to the right immediately under the selected major subclass, selecting the first appearing title to include an additional characteristic of the invention.

Referring back to the schedule of titles in the *Manual of Classification,* there are two subclasses indented one-dot: "From nitric acid" and "From organic compound." The latter is more pertinent in this search; therefore, the desired subclass is indented under this subclass.

STEP 8. Proceed until the most specific minor subclass is reached whose title and definition indicate the inclusion of the invention in question and then inspect the art composing it.

By this process of elimination, we eventually reach subclass 19: Ethanol. The definitions indicate that indeed this is the proper subclass for searching. If this search were being conducted at the public search room, it would be a good idea to have this field of search confirmed by an examiner as discussed in a previous section.

In this search, the proper subclass was identified. Had difficulties been encountered, one would have proceeded with the next step.

STEP 9. If this subclass contains nothing pertinent, search the more generic subclass under which it is indented, the miscellaneous subclasses, and those above it, the titles of which indicate the possibility that they may include the subject matter sought.

STEP 10. If the pertinent art still is not found, search other possible classes noted in step 2 in the same way.

Failure to find any pertinent art usually indicates that the proper place in the classification has not been located. The classifiers engaged in creating new classifications and maintaining the existing ones are always available for consultation in outlining fields of search.

PROFESSIONAL SEARCHERS

There are a number of individuals, firms, and companies which will perform patent searches for a fee. Some firms have access to the previously discussed Patent Office database, APS, which they use for searching. The quality of a search depends on the qualifications of the searcher, his access to databases, and the time allowed for the search. When looking for a searcher, it is good to discuss each of these.

Two companies which use the APS database are RWS Information, Inc., located at 1919 S. Eads Street, Arlington, VA 22202, and Rapid Patents, 1921 Jefferson Davis Highway, Arlington, VA 22202. Both companies offer several

CLASS 203 DISTILLATION: PROCESSES, SEPARATORY

1	WITH MEASURING, TESTING OR INSPECTING
2	.Of temperature or pressure
3	.Of concentration
4	INCLUDING PURGING OF THE SYSTEM
5	SEPARATING ISOTOPES OR TAUTOMERS
6	ADDITION OF MATERIAL TO DISTILLAND TO INHIBIT OR PREVENT REACTION OR TO STABILIZE
7	.For scale inhibiting or corrosion preventing
8	.For inhibiting or preventing a polymerization reaction
9	..Of unsaturated hydrocarbon
10	WATER PURIFICATION ONLY
11	.Under pressure or vacuum
→ 12	DISTILLING TO SEPARATE OR REMOVE ONLY WATER
13	.From nitric acid
14	.From organic compound
15	..Organic acid
16	...Acetic
17	..Aldehyde or ketone
18	..Alcohol
19	...Ethanol
20	INCLUDING DEFOAMING OR INHIBITING FOAM
21	AND RECOVERING HEAT BY INDIRECT HEAT EXCHANGE
22	.Utilizing recovered heat for heating feed
23	..Distillation residue as heat source
24	..Compressed vapor as heat source
25	.Utilizing recovered heat for heating the distillation zone
26	..Compressed vapor as heat source
27	.Utilizing recovered heat in subsequent step in process
28	WITH CHEMICAL REACTION
29	.Including step of adding catalyst or reacting material
30	.For polymerizing unwanted component
31	..Oxidizing material
32	..Reducing material
33	..Inorganic salt containing oxygen in the anion
34	..Acid
35	..Phosphoric or sulfuric acid
36	..Alkaline oxide or hydroxide
37	..Alkali metal hydroxide
38	..Organic material
39	WITH DISPARATE PHYSICAL SEPARATION
40	.Of entrained particles from a vapor or gas
41	.Utilizing solid sorbent
42	.Utilizing liquid sorption of component from gas or vapor
43	.Utilizing liquid-liquid extracting of distillation product
44	..Of distillate
45	...And distilling raffinate phase
46	...And distilling extractant phase
47	.Utilizing removing solid from liquid
48	..By crystallizing
49	CONVECTIVE DISTILLATION WITH NORMALLY GASEOUS MEDIUM, E.G., AIR
50	ADDING MATERIAL TO DISTILLAND EXCEPT WATER OR STEAM PER SE
51	.At least two materials
52	..Mixtures of hydrocarbons
53	..One material is water
54	...A second material is aldehyde or ketone
55	...A second material is an alcohol
56	..One additive an alcohol or ether
57	.Organic compound
58	..Heterocyclic
59	..Amine
60	..Ester
61	..Acid

62	..Aldehyde or ketone
63	..Alcohol or ether
64	...Polyhydroxy alcohol or ether derivative thereof
65	...Hydroxy aromatic (e.g., Phenol)
66	...Methanol
67	...Halogenated hydrocarbon
68	..Hydrocarbon
69	...Aromatic
70	...Acyclic
71	PLURAL DISTILLATIONS PERFORMED ON SAME MATERIAL
72	.One a filming distillation
73	.One a distillation under positive pressure or vacuum
74	..Distillation of initial distillate
75	...And return of distillation product to a previous distillation zone
76	....Including the addition of water or steam
77	...Initial distillation under positive pressure or vacuum
78	...And returning distillation product to a previous distillation zone
79	....Including the addition of water or steam
80	...Initial distillation under positive pressure or vacuum
81	.Distillation of initial distillate
82	..And returning distillation product to a previous distillation zone
83	..Including the addition of water or steam
84	.And returning distillation product to a previous distillation zone
85	.Including the addition of water or steam
86	DISTILLATION IN APPARATUS OR ELEMENT OF SPECIFIC MATERIAL
87	WITH FRACTIONAL CONDENSATION OF VAPOR OUTSIDE STILL
88	FLASH VAPORIZATION OF DISTILLAND
89	FILMING OF DISTILLAND FOR VAPORIZATION
90	SPRAYING OF DISTILLAND INTO VAPORIZATION ZONE
91	VAPORIZATION ZONE UNDER POSITIVE PRESSURE OR VACUUM
92	.Including the addition of water or steam
93	..And returning product of distillation step to distillation zone
94	.And returning product of distillation step to distillation zone
95	INCLUDING ADDITION OF WATER OR STEAM
96	.To distillation column
97	.And returning product of distillation step to distillation zone
98	AND RETURNING PRODUCT OF DISTILLATION STEP TO DISTILLATION ZONE
99	MISCELLANEOUS SEPARATORY
	CROSS-REFERENCE ART COLLECTION
100	PARTICULAR TYPE OF HEATING
	DIGESTS
DIG 1	Solar still
DIG 2	Laboratory distillation
DIG 3	Acrylonitrile
DIG 4	Heat pump
DIG 5	Human waste
DIG 6	Reactor-distillation
DIG 7	Start up or shut down
DIG 8	Waste heat
DIG 9	Plural feed
DIG 10	Vinyl acetate
DIG 11	Batch distillation

Schedule of subclasses for Class 203 (from the *Manual of Classification*).

types of searches. Preliminary patent searches are about two hundred dollars and are delivered within two days.

There are a number of Washington-based law firms which will perform searches, for example, Alan Faber, 3452 South Stafford Street, Arlington, VA 22206. Their hourly rate should be in the fifty-dollar-per-hour range.

OBTAINING UNITED STATES PATENT COPIES

If one has access to either the public search room at the Patent Office or to a depository library, he can make copies since patents are not subject to copyright. Both places have token-operated photocopy equipment.

Copies can be ordered by mail from the U.S. Patent and Trademark Office, Washington, D.C. 20231. Copies are three dollars per copy. Information can be obtained from the Patent Office by phone.

There are also a number of private companies that specialize in procurement of patents. While patent copies from these companies may be a bit more expensive than those obtained from the Patent Office, their services are more personalized, and their speed of delivery much quicker. RWS Information and Rapid Patent, addresses given in the previous section, are examples.

SEARCHING FOREIGN PATENTS

As previously discussed, most foreign countries classify their patent documents by the International Patent Classification System.

Since foreign patent offices do not classify documents by the United States Patent Classification System, the Classification Division must do so before these documents can be placed in the United States classified files. This is a tremendous task, and many non–English language documents have not been classified for the United States classified files. Foreign documents that have been classified by the United States system are filed in the examiners' search files; they are not filed in the classified files in the public search room.

It is sometimes necessary to search foreign patents and publications in addition to United States references, as when conducting invalidity and infringement searches. There are two European search facilities available to the public which will undertake a confidential search upon request without filing a patent application, provided one pays their fee. These two facilities are the European Patent Office and the Swedish Patent Office. Both facilities will accept search requests from anyone, conduct an international search, and report the results in English. While these searches are expensive, they are quite good.

To follow recently published patent documents from foreign countries, one can follow the gazette published by each office, or one can subscribe to a commercial service which prescreens and abstracts issuing documents of different countries. Essentially all countries that publish patent documents publish a gazette abstracting these documents. These gazettes are offered by subscription, or one can read them in a library. Copies of gazettes for most countries are available at the United States Patent Office. To use the gazettes effectively, one must be familiar with the International Patent Classification System; otherwise, not a lot of information can be obtained unless one is looking for a specific document. Even one familiar with the classification system is faced with a translation problem.

A number of commercial companies prescreen, classify, and publish in English an abstract of recently published patent documents from foreign countries. Leaders in this area include Chemical Abstract, Derwent Patent Information Services, and International Patent Documentation Center, all discussed previously in this chapter.

Searches by the European Patent Office

The European Patent Office (EPO) was first established in January 1978. At that time the International Patents Institute ceased to exist as an independent organization, and its facilities were incorporated into the European Patent Office. The facilities are located in Munich, Germany.

The EPO was established with two objectives: to search and examine patent applications of member countries of the European Patent Convention and to function as an international searching authority. In its function as an international search authority, the EPO provides three types of services to the public: standard search, special search, and on-line information.

The standard search is identical to the official search provided for under the European Patent Convention. The search encompasses the whole of the documentation in the parts of the EPO classification system relevant to the subject matter of the invention. The search is based on the invention as defined in the claims. If no claims are present in the documents as submitted to be searched, the EPO prefers that the client indicate what he considers to be his invention by a written summary. For example, this situation may exist with a United Kingdom provisional specification.

A standard search may be requested by an individual or any legal entity on inventions for which a patent application has been filed or a patent has been granted. This type of search may be useful when making a decision on whether to file a foreign application or when making a decision on initiation of infringement.

Special searches by the EPO are designed to meet specific requirements of a client; these searches are not based on filed patent applications or issued patents. The client submits instructions on the type of information in which he is interested, and EPO personnel structure the search around this interest. The fees for special searches are determined in each case based upon the amount of work necessary. This type of search can be very useful before filing a patent application, particularly, a foreign patent.

Approximately 20% of all patent applications are rejected during the examination phase as being anticipated by prior art. If sufficient time is available, a prior art search should be made in every case before preparing a patent application. It is much less expensive to conduct a patentability search than to prepare a patent application. This is particularly true of foreign applications. Foreign applications are expensive because of legal fees, filing fees, etc. If the prior art search produces art which would probably preclude the claims, the client is spared the expense of preparing the application. If the subject matter is not anticipated by prior art, the pertinent prior art will provide background information for preparing the application.

With United States applications, it is always beneficial to make the examiner aware of any foreign prior art. The United States classification files are not well filed with foreign prior art. If the patent issues and is commercially successful, competitors will consider infringement and will have foreign searches conducted in an attempt to invalidate the patent. It is a fact that if only the Patent Office prior art record is presented during the infringement proceeding, the court usually rules in favor of the patentee; if additional pertinent prior art is presented, the court is more likely to rule in favor of the defendant.

On the other hand, if you are a defendant in an infringement proceeding, you need to know the identity of documents which can lead to the infringed patent being declared invalid. In such cases, the EPO can perform a very thorough search.

When establishing long-term research or manufacturing policy, a company needs to be aware of the state of the prior art; the company should be aware of all documents relevant to the particular field. From a research standpoint, it is less expensive to search and distinguish prior art than to experiment; from a production standpoint, it is not wise to invest in a production facility for a product that might infringe another patent.

Searches by the Swedish Patent Office

The Swedish Patent Office, located in Stockholm, offers a number of search services. As early as 1947, this patent office offered novelty searches to members of the public even if they had not filed applications within the office. During recent years, these services have been greatly expanded, and now the Swedish Patent Office offers several searches, including novelty searches, state of the art searches, and general information searches.

The search request can be in English, French, or German as well as the

Scandinavian languages. No special requirements exist for a search request; the client defines the subject matter as precisely as possible and indicates the type of search or information in which he is interested. The search report itself is a list of the countries and the technical fields searched and a list of documents considered relevant. If requested, photocopies of cited documents are sent to the client at his expense. Normally the search report is complete within 1 to 4 weeks.

The Japanese Patent Office

During recent years Japan has become very active in research, particularly chemical and electronic research. As a result, a lot of technical information is disclosed in Japanese patent documents. The Patent Office of the Japanese Government (POJG) publishes four types of patent documents:

1. KOKAI patents (unexamined)
2. KOKOKU patents (examined)
3. KOKAI utility models (unexamined)
4. KOKOKU utility models (examined)

The original gazettes in Japanese for all of the above publications are sent by the POJG to the United States Patent Office. At present, KOKAI (unexamined) patents are kept on shelves in numerical order for public access in the scientific library; other booklets are kept in storage because of space considerations. Also, the U.S. Patent Office has microfilm of all four original gazettes available in the scientific library. These can be read and copied from facilities available there.

The POJG provides English abstracts of various Japanese patent documents. These publications contain bibliographic information and English abstracts of approximately 150 words. However, these English abstracts are made only from KOKAI (unexamined) patent applications in certain technical fields; coverage is about 50% of the KOKAI patents.

These abstracts are published in booklets containing 500 abstracts per booklet; they are divided into the four groups: general and mechanical, chemical, physical, and electrical. Annually, about 25 booklets are published, and these issue about two months after publication of KOKAI patent applications. These booklets are kept on shelves in the scientific library for public use. These booklets can also be purchased from the Japan Institute of Invention and Innovation.

For the KOKAI (unexamined) patent applications, the POJG produces three kinds of indexes: numerical order (JPIN), applicant order (JPIA), and classification order (JPIC). Each index is marked for English abstracts, if any, showing abstract section, year, volume number, and page, so that one can easily consult the English abstract for documents of interest.

There are at least two Japanese institutions from which one can obtain information. The Japanese Institute of Invention and Innovation (JIII), as previously stated, is a semiofficial institute of the Japanese government which publishes the patent publications PAG and JPI under the guidance of the POJG. In addition, the JIII offers several kinds of searches by request. These searches are available in Japanese and English and may be requested from overseas.

The Japan Patent Information Center (JAPATIC) is another semiofficial institute of the Japanese government which provides computer retrieval services for patent information under the guidance of the POJG. These services are best utilized through a representative in Japan.

OBTAINING FOREIGN PATENT COPIES

The United States Patent Office maintains a fairly extensive foreign patent file for most foreign countries. Foreign patents are numerically arranged for each country on either paper or microfilm in the scientific library on the second floor.

Foreign patent copies can also be ordered from the national patent office publishing the document. The addresses of certain foreign patent offices are listed at the end of this chapter. In addition, addresses of foreign patent offices, along with prices for copies of patents, are listed in the Introduction to the semiannual volumes of *Chemical Abstracts*.

The European Patent Office will supply photocopies of any of the documents in its collection. As far as the patent documents are concerned, this documentation includes a large number of countries. The time required for the supply of photocopies is usually short. Accessible documents, in particular patent documents, are usually supplied within a few days. The address of the European Patent Office is given in an earlier section.

One can also obtain copies from private companies specializing in procurement of patent copies. These tend to give quicker and more personalized service than most national patent offices. For foreign patents, the best company in this area is probably Derwent, whose address was given in the previous section.

Another source for foreign patent copies is the British Library, located at 128 Theobalds Road, London WCIX8RP, England. One can use ORBIT or DIALOG to order.

ADDRESSES OF FOREIGN PATENT OFFICES

Austria   OSTERREICHISCHES PATENTAMT
                               Kohlmarkt 8-10
                               A-1014 Wien [Vienna]

Belgium	SERVICE DE LA PROPRIETE INDUSTRIELLE ET COMMERCIALE Ministère des Affaires économiques 24-26, rue J.A. de Mot *B-1040 Bruxelles [Brussels]*
Denmark	DIREKTORATET FOR PATENT-OG VAREMAERKEVAESENET Nyropsgade 45 *DK-1602 Kobenhavn V [Copenhagen]*
Finland	PATENTTI-JA REKISTERIHALLITUS PATENT-OCH REGISTERSTYRELSEN Bulevardi 21 *SF-00180 Helsinki 18*
France	INSTITUT NATIONAL DE LA PROPRIETE INDUSTRIELLE 24 bis, rue de Leningrad *F-75008 Paris*
Germany	DEUTSCHES PATENTAMT Zwiebruckenstraße 12 *D-8000 München 2 [Munich]*
Greece	DIRECTORATE FOR COMMERCIAL AND INDUSTRIAL PROPERTY Ministry of Trade Kaningos Square *Athens*
Ireland	INDUSTRIAL AND COMMERCIAL REGISTRATION OFFICE 45 Merrion Square *Dublin-2*
Italy	L'UFFICIO CENTRALE BREVETTI Via Molise 19 *I-00187 Roma [Rome]*
Japan	TOKKYOCHO 1-3-1 Kasumigaseki Chiyoda-ku *Tokyo*
Luxembourg	SERVICE DE LA PROPRIETE INDUSTRIELLE Ministère de l'Economie nationale et des Classes moyennes 19-21, boulevard Royal B.P. 97 *Luxembourg*

Netherlands	OCTROOIRAAD Patentlaan 2 *Rijswijk (ZH)*
Norway	STYRET FOR DET INDUSTRIELLE RETTSVERN Middelthunsgate 15b/17 Postboks 8160 Dep. *N-Oslo 1*
Portugal	INSTITUTO NACIONAL DA PROPRIEDADE INDUSTRIAL Campo das Cebolas *Lisboa [Lisbon]*
Spain	REGISTRO DE LA PROPIEDAD INDUSTRIAL Avda. del Generalisimo, 59 *Madrid 16*
Sweden	KUNGLIGA PATENT-OCH REGISTRER- INGSVERKET Valhallavagen 136 P.O. Box 5055 *S-102 42 Stockholm*
Switzerland	EIDGENOSSISCHES AMT FUR GEISTIGES EIGENTUM Eschmannstraße 2 *CH-3003 Bern*
United Kingdom	UNITED KINGDOM PATENT OFFICE 25 Southampton Buildings Chancery Lane *London WC2A 1AT*
Russia	RUSSIAN STATE COMMITTEE FOR INVENTIONS AND DISCOVERIES Malyi Tcherkasski Pereulok *2/6 Moscow*

Chapter 16

REQUIREMENTS OF A REFERENCE

INTRODUCTION

When an inventor desires patent protection, he must submit appropriate information on the invention to the patent office in the country in which he desires protection. To determine if the subject matter meets the conditions for patentability, that is, it is novel and nonobvious in the United States, the patent examiner does a "state-of-the-art search" for the subject matter at that point in time. The state-of-the-art search is done by analyzing prior art reference materials.

A reference is any item of evidence which indicates that a claimed invention existed in the prior art at a given point in time. For the item of evidence to preclude patentability by United States patent statutes, it must contain sufficient information to enable a person of ordinary skill in the pertinent technology to make and use the claimed invention without performing extensive experimentation. This is the so-called "enabling disclosure" requirement.

As discussed in Chapter 3, 35 U.S.C. Sect. 102(a) provides that a person is not entitled to a patent if the invention was known or used by others in this country before the invention thereof by the applicant. According to this statute, prior knowledge of the invention is sufficient evidence to preclude patentability. However, establishing prior knowledge by any method other than documentary evidence is very difficult from a legal standpoint. Since the patent office is not in a position to obtain evidence other than documentary evidence, rejections other than those based on documentary evidence are extremely rare. Domestic patents, foreign patents, and printed publications are the reference forms most often cited as prior art as evidence of prior knowledge by others.

With regard to printed matter, 35 USC Sect. 102(a) provides that a person is not entitled to a patent if the invention was patented or described in a printed publication in this or a foreign country before the invention thereof by the applicant. This section specifically identifies three forms of prior art evidence: patents, patent applications, and printed publications. Because of the inherent authenticity of these documents, the contents of these documents are generally accepted as prior art knowledge per se, rather than evidence of prior art.

Patents are obviously printed publications, but, in patent law, issued patents are distinguished from certain other printed publications such as books, technical and scientific journals, etc., because these are treated differently by the patent statutes. These differ mainly as to when they are available as prior art references. The effective reference date of an issued United States patent is the filing date of the application from which the patent issued; the effective reference date of a foreign patent is its issue date (date of vested rights); the effective reference date of a printed publication is the date of publication, or more exactly the date it is available to the public.

THE ENABLING DISCLOSURE REQUIREMENT

In order for an invention to be patentable in the United States, it must be novel, it must be nonobvious to those skilled in the pertinent art, and it must have utility. Section 102 of the Patent Code specifies the novelty requirements for patentability along with certain conditions which may cause a loss of right to patent. Section 103 specifies the nonobvious requirement for patentability: the invention must be sufficiently different from what has been used or described in the past that it would not be obvious to a person ordinarily skilled in the pertinent art. Section 101 specifies the utility requirement for patentability. The novelty and nonobviousness requirements relate to prior art.

Over many years of case law, a concept commonly referred to as the "enabling disclosure" requirement has evolved which sets forth the minimum level which any reference must meet to preclude patentability. In essence, for a prior art reference under 35 U.S.C. Sect. 102 to be anticipatory of a claimed invention, the reference must contain information sufficient to enable a person of ordinary skill in the pertinent art at a point in time preceding the applicant's invention date to make and use the claimed subject matter without performing extensive experimentation. A Section 102 rejection usually involves a single reference. The "enabling disclosure" test applies equally to references under Section 103. With Section 103 rejections, two or more references can be combined to make the invention obvious to a person of ordinary skill in the pertinent art at that point in time.

When determining if a reference discloses (anticipates or makes obvious) the invention, one must make the following assessment of the reference(s) involved:

1. What type of information does the reference contain?
2. To what type of person would this information be pertinent?
3. At what point in time was this information available?

The enabling disclosure requirement varies somewhat depending upon the technology involved and upon the skill associated with the "ordinary man" in the technology. Because of these variations, this edition does not attempt to set forth particulars relating to prior art evidence in different technologies. The following generalizations can be made with regard to Sections 102 and 103, however.

For a Section 102 rejection to be proper, the prior art reference must both describe and illustrate the subject matter of the claim "within four corners"; the reference must clearly anticipate the claim. Most courts have interpreted that a Section 102 rejection must meet this requirement, and this is the interpretation followed by the Patent Office. Therefore, a Section 102 rejection usually involves single reference. If the examiner uses a combination of references for a 102 rejection, this is most often an improper rejection.

As indicated above, a reference cited in a Section 102 rejection must both describe and illustrate the subject matter of the application. The reference may do this in varying degrees, however. At one extreme, the reference may disclose the entire concept of the invention as well as the structure of the invention, thereby anticipating all claims. This rejection can be overcome only by swearing behind the reference date by a Rule 131 affidavit.

On the other hand, the reference may only anticipate one or two of the claims of the application and not disclose the entire inventive concept. In this case, it is a matter of redrafting the claims by amendment to overcome the reference. This situation is quite common since claims are often submitted which are drafted as broad as prior art will permit, and the broader claims are rejected whereas the narrower claims are allowed.

As previously stated, the "enabling disclosure" test applies equally to Section 103. A Section 103 reference or combination of references must render the claimed invention obvious to a person of ordinary skill in the art at a period just prior to the invention date of the subject matter.

When the examiner is unable to find a reference that identically discloses or describes the subject matter as required in Section 102 rejections, he may find one or more references which make the subject matter obvious to a person skilled in the relevant art. As discussed in Chapter 4, Section 103 states that the subject matter must be nonobvious to a person skilled in the relevant art at the time of the invention for a valid patent to issue. For a Section 103 rejection, the examiner often uses a combination of two or more references.

When making a Section 103 rejection, the examiner must state the differences between the claimed subject matter and the cited references, must indicate the proposed modifications of the cited references required to meet the claimed subject matter, and must explain why such modifications would be obvious to a person skilled in the art at the time of the invention. This requirement is set forth in the *Manual of Patent Examining Procedure*, Section 706.02. This information can be very useful to the applicant in preparing his response to the rejection.

PATENTS AND PATENT
APPLICATIONS AS REFERENCES

Subsection 102(a) and subsection(b) provide that "a person shall be entitled to a patent unless . . . the invention was patented in this or a foreign country." Subsection 102(a) applies to patent documents having a reference date prior to the inventor's date of invention; subsection 102(b) applies to patent documents which issued more than one year prior to the applicant's United States filing date.

As a general rule, pertinent issued United States patents comply with the disclosure requirements discussed in the previous section. By statute, a United States patent must "contain a written description of the invention and the manner and process of making and using it, in such full, clear, concise, and exact terms as to enable any person skilled in the art to which it pertains, or with which it is most nearly connected, to make and use the same." Therefore, when a United States patent document describes the same "inventive concept" as contained in an application, the patent usually has a disclosure sufficient to anticipate the application. However, while a United States patent reference may by definition contain an enabling disclosure of the subject matter disclosed in the patent as a reference, this does not necessarily mean that the patent also contains an enabling disclosure of subject matter claimed in an application. Obviously, they must relate to the same inventive concept.

As a general rule, foreign patent documents do not contain a detailed disclosure as is the case with United States patent documents. Most foreign countries have a minimal disclosure requirement, and patent documents usually do not describe in detail the subject matter which the document is claiming. For this reason, foreign patent documents are subject to various interpretations as to what they exactly disclose.

By statute, United States patent applications are filed and prosecuted under conditions of secrecy. For this reason, patent applications cannot be relied upon by the Patent Office as evidence of prior knowledge. Likewise, abandoned or unissued United States patent applications are normally maintained in confidence and are thus not available as prior art evidence.

Patent documents have the following effective reference dates:

1. The effective date of an issued United States patent under sections 102(a), 102(e), and 102(g) is the filing date of the application.

2. The effective date of an issued United States patent under section 102(b) is its issue date.

3. The effective reference date of a foreign patent is its issue date, that is, the date on which vested rights are granted to the applicant.

4. Pending, abandoned, unissued United States applications normally are not available as prior art references.

As discussed in previous chapters, many foreign countries publish patent applications during the course of prosecution. Publication of applications under these systems does not actually vest patent rights, but serves notice to infringers that should patent rights ultimately be vested to the applicant, damages for past infringement can be collected. As a general rule, these published applications have not served as a prior art reference; that is, foreign patent documents were available as prior art on the date on which vested patent rights were granted. However, the Patent Board of Appeals recently ruled that publications of a Japanese patent application constituted the patenting of the claimed invention within section 102(d). Therefore, there may be some rethinking in this area in the future.

PRINTED PUBLICATIONS AS REFERENCES

Subsection 102(a) and subsection 102(b) provide that "a person shall be entitled to a patent unless ... the invention was described in a printed publication in this or a foreign country." Subsection 102(a) applies to printed publications having an effective reference date prior to the applicant's date of invention; subsection 102(b) applies to printed applications having an effective reference date more than one year prior to the applicant's actual United States filing date.

In general, printed publications include any publication where the text is fixed or impressed on pages and would include text books, scientific periodicals and journals, trade magazines, technical manuals, catalogs, government publications, etc.

The effective reference date of a printed publication is the date on which it is available to the public, that is, the date on which the publisher releases the publication for public distribution. Take for example a scientific periodical. These often have an issue date printed on the cover, but, this is not the effective reference date. The effective reference date of the contents is the date on which the publication becomes available to the public.

As a general rule, technical publications tend to be directed toward broader scientific audiences and are therefore often written in general terms. In this regard, printed publications usually do not meet the enabling disclosure requirement previously discussed. Most often, this type publication does not contain information sufficient to enable a person to make and use the disclosed subject matter without experimentation. This is a generalization, however, and each publication is subject to interpretation as to exactly what it discloses.

DUTY OF DISCLOSURE OF PRIOR ART

There is a duty of disclosure of pertinent prior art on the part of the inventor, the attorney, or the agent who prepares the application, and anyone involved in the prosecution of the application or research of the invention. Failure to make such disclosure may result in a fraudulent patent. The duty of disclosure is set forth in 37 CFR 1.56 cited below in pertinent part.

> (a) A duty of candor and good faith toward the Patent and Trademark Office rests on the inventor, on each attorney or agent who prepares or prosecutes the application and on every other individual who is substantively involved in the preparation or prosecution of the application and who is associated with the inventor, with the assignee or with anyone to whom there is an obligation to assign the application. All such individuals have a duty to disclose to the Office information they are aware of which is material to the examination of the application. Such information is material where there is a substantial likelihood that a reasonable examiner would consider it important in deciding whether to allow the application to issue as a patent. The duty is commensurate with the degree of involvement in the preparation or prosecution of the application.
>
> (b) Disclosures pursuant to this section may be made to the Office through an attorney or agent having responsibility for the preparation or prosecution of the application or through an inventor who is acting in his own behalf. Disclosure to such an attorney, agent or inventor shall satisfy the duty, with respect to the information disclosed, of any other individual. Such an attorney, agent or inventor has no duty to transmit information which is not material to the examination of the application.

The two major sources of prior art are the literature search which the inventor did for research purposes, and the patentability search that the patent practitioner conducts to determine patentability. From this list, the decision must be made about which references to disclose and which not to disclose.

Once the decision has been made that a reference is material and should be cited, next one must choose the best method of informing the examiner about the prior art. There are two methods the reference can be included in the specification or the reference can be made part of the prior art statement. The prior art statement is provided for in 37 CFR Sects. 1.97, 1.98, and 1.99. Section 1.97 sets forth the format and the time for filing the prior art statement and is cited below.

> (a) As a means of complying with the duty of disclosure set forth in sect. 1.56, applicants are encouraged to file a prior art statement at the time of filing the application or within three months thereafter. The statement may either be separate from the specification or may be incorporated therein.
>
> (b) The statement shall serve as a representation that the prior art listed

therein includes, in the opinion of the person filing it, the closest prior art of which that person is aware; the statement shall not be construed as a representation that a search has been made or that no better art exists.

Rule 98 sets forth the content of the prior art statement and is cited below.

(a) Any statement filed under sect. 1.97 or sect. 1.99 shall include: (1) A listing of patents, publications or other information and (2) a concise explanation of the relevance of each listed item. The statement shall be accompanied by a copy of each listed patent or publication or other item of information in written form or of at least the portions thereof considered by the person filing the statement to be pertinent.

(b) When two or more patents or publications considered material are substantially identical, a copy of a representative one may be included in the statement and others merely listed. A translation of the pertinent portions of foreign language patents or publications considered material should be transmitted if an existing translation is readily available to the applicant.

Rule 99 provides for updating the prior art statement and is cited below.

If prior to issuance of a patent an applicant, pursuant to his duty of disclosure under sect. 1.56, wishes to bring to the attention of the Office additional patents, publications or other information not previously submitted, the additional information should be submitted to the Office with reasonable promptness. It may be included in a supplemental prior art statement or may be incorporated with other communications to be considered by the examiner. Any transmittal of additional information shall be accompanied by explanations of relevance and by copies in accordance with the requirements of sect. 1.98.

The Patent Office requests that applicants use form PTO-1449, "List of Prior Art Cited By Applicant," when submitting a prior art statement under 37 CFR 1.97–1.99. Each reference listed on this statement, when considered by the examiner, will be printed on the issued patent as a prior art reference. This form can be obtained from the Patent Office.

DEFENSIVE PUBLICATIONS

A number of companies file patent applications strictly for defensive purposes. This is done to preclude competitors from obtaining patent protection in a certain area of development; the issued patent serves as prior art against the competitor's patent application. While this approach can be effective in keeping an area of development open, it does have disadvantages. First, it is expensive, and second, it ties up patent examiners in examining applications for which no commercial development is likely.

Being aware of the above approach and problems associated with it, the Patent Office maintains a Defensive Publication Program. By this program, if an applicant waives his rights to an enforceable patent on a submitted application, the Patent Office will publish an abstract on the disclosure of the application and make the application publicly available as prior art. Patent Office Rule 139 relates to the Defensive Publication Program and is cited below.

> An applicant may waive his rights to an enforceable patent based on a pending patent application by filing in the Patent and Trademark Office a written waiver of patent rights, a consent to the publication of an abstract, and authorization to open the complete application to inspection by the general public, and a declaration of abandonment, signed by the applicant and the assignee of record or by the attorney or agent of record.

Chapter 17

PATENT RIGHTS

INTRODUCTION

The United States patent grants to the patentee the right to exclude others from making, using, or selling the patented invention throughout the United States, its territories, and its possessions for the life of the patent. The key phrase in this expression is the "right to exclude." The patent does not grant the patentee the right to make, use, or sell the invention, but the right to exclude others from commercial exploitation of the invention while the patentee may himself exploit it. The exclusionary nature of the patent right is discussed in the first section of this chapter. If another without proper authority makes, uses, or sells the patented invention within the territory of the United States, he infringes the patent. When this happens, the patentee may sue for relief in federal court. He may ask for an injunction to prevent continuation of the infringement, and he may ask for damages caused by the infringement. Infringement is a civil action, not a criminal action, and suits are initiated in federal court in the district where the infringing activity occurred.

The Patent Office has no jurisdiction over questions of infringement of patents. When the patent issues, it is presumed valid. During infringement litigation, however, the defendant may raise the question of the validity of the patent, and the court determines the question. The second section of this chapter is concerned with infringement and the defenses for infringement.

Another section in this chapter deals with the rights of the employee and the employer. Most of the research done in the United States occurs in an employee-employer relationship. This section discusses the fact that while the title to the patent must be in the name of the inventor, the title is subject to vestment to the employer in cases of employed inventors.

EXCLUSIONARY NATURE

As previously stated, the patent does not grant to the patentee the right to make, use, or sell the patented invention, but the right to exclude others from doing so. Section 271 of the Patent Code relates to others who without authority make, use, or sell the patented invention; it is cited below in pertinent part:

> (a) Except as otherwise provided in this title, whoever without authority makes, uses or sells any patented invention, within the United States during the term of the patent therefor, infringes the patent.

Since the patent does not grant the right to make, use, or sell the invention, the patentee's right to do so depends upon the rights of others and any laws applicable. If the making, using, or selling of the patented invention violates any general laws applicable, the patentee is not authorized to do so. If making, using, or selling of the invention infringes the prior patent rights of others, the patentee is not authorized to do so. As later discussed, in examining applications the Patent Office does not make an absolute determination whether the invention sought to be patented infringes any other prior patent. In addition, the patent rights do not give the patentee authority to violate any of the federal antitrust laws. Usually, however, there is nothing that prevents the patentee from making, using, or selling his invention unless his activity infringes another patent that is still in force.

The patent itself creates no marketable product and puts no money in the bank. With proper commercial exploitation, however, the patent may become a very valuable property. If the patent encompasses marketable subject matter, the patent allows the patentee to exclude others from commercial exploitation of the invention while the patentee may engage in such exploitation. It is this exclusionary right that gives incentive for research and development, and it is this right that sometimes makes the patent a very valuable property.

Section 261 of the Patent Code states in pertinent part:

> Subject to the provisions of this title, a patent shall have the attributes of personal property.... In accordance with this statute, patent rights may be sold (assigned), mortgaged, or licensed to others.

An assignment involves a total transfer of the patent rights to the assignee for the life of the patent. The assignee thereafter has the right to exclude others from making, using, or selling the patent invention. If any of these rights are withheld, the agreement constitutes a license rather than an assignment.

A license is something less than total transfer of property; the patentee still has title to the invention. A license is, in effect, an agreement to allow another to infringe the patent. A license may be exclusive or nonexclusive. An exclusive license gives another the exclusive right to make, use, or sell the invention but does not transfer the title, whereas a nonexclusive license may allow several persons to share in the field.

Large companies usually exploit their patent rights by manufacturing and selling the patented machine, article of manufacture, or composition of matter, or they use their patented process to produce sellable goods. By these rights, the patentee (or assignee) can exclude others from simultaneous commercial exploitation of the invention. This greatly increases the chances of commercial success of the invention.

For the individual inventor, however, this avenue might not be available. It often takes large amounts of capital to develop and market a product. The individual inventor may therefore choose either to sell his patent or to license certain rights to his patent.

INFRINGEMENT

If anyone without authority makes, uses, or sells a patented invention within the United States during the life of the patent, he infringes the patent. The statutory provision for infringement is set forth in Section 271 of the Patent Code cited below.

> (a) Except as otherwise provided in this title, whoever without authority makes, uses or sells any patented invention, within the United States during the term of the patent therefor, infringes the patent.
>
> (b) Whoever actively induces infringement of a patent shall be liable as an infringer.
>
> (c) Whoever sells a component of a patented machine, manufacture, combination or composition, or a material or apparatus for use in practicing a patented process constituting a material part of the invention, knowing the same to be especially made or especially adapted for use in an infringement of such patent, and not a staple article or commodity of commerce suitable for substantial noninfringing use, shall be liable as a contributory infringer.
>
> (d) No patent owner otherwise entitled to relief for infringement or contributory infringement of a patent shall be denied relief or deemed guilty of misuse or illegal extension of the patent right by reason of his having done one or more of the following: (1) derived revenue from acts which if performed by another without his consent would constitute contributory infringement of the patent; (2) licensed or authorized another to perform acts which if performed without his consent would constitute contributory infringement of the patent; (3) sought to enforce his patent rights against infringement or contributory infringement.

By this statute there are three types of infringement: direct, inductive, and contributory. Direct infringement involves directly making, using, or selling patented subject matter without authority. For this activity, the infringer is liable as a direct infringer.

Subsection (b) of Section 271 relates to inductive type infringement. This involves inducing another to infringe a patent. For example, if one sold to another ingredients for a patented process and instructed the buyer on how to use the ingredients in this process, the seller of the ingredients would be liable as an infringer by inducement. In other words, the seller commits an act that induces another to infringe. The buyer that used the process directly would be liable as a direct infringer.

Contributory type of infringement involves selling a component of a patented assembly or a material for a patented process to one other than the patentee of the assembly or process when these are made especially for the patented machine or process. For example, if a company makes an unpatented component for express use in a patented machine and this component has no other use than in this patented machine, selling of this component to one other than the patentee would constitute contributory infringement. Note however, that this component must be especially made for, or adapted for, the patented machine and have no suitable use other than in the patented machine.

Since it is the claims that define the invention, questions of infringement are determined primarily by the language of the claims. If the activity of the alleged infringement falls within the scope of the claim, the claim is being infringed; if the activity does not fall within the scope of the claim, the claim is not being infringed. Certain judicial doctrines have expanded the scope of this somewhat, however. Two such doctrines are the doctrine of equivalents and the doctrine of colorable deviation.

As an example of the doctrine of equivalents, consider a patent with a claim consisting of elements W, X, Y, and Z. If another is making, using, or selling subject matter which omits one of these elements, he is not infringing that claim by strict definition. However, if he is substituting another element in the place of the omitted element which is equivalent to the omitted element, the courts may rule that he is infringing the claim by the doctrine of equivalents. This occurs most often with mechanical cases.

The doctrine of colorable deviation occurs most often in chemical technology and may be appropriate when someone is operating just outside of a range set by a claim in an attempt to avoid infringing the claim. Consider, for example, a patented chemical process with a claim setting temperature limits of 40–60 degrees. If another is operating this process just outside this temperature range, say at 38 degrees, and the process is less efficient at this temperature, the courts may rule that the patent is being infringed by the doctrine of colorable deviation.

The remedy for infringement is set forth in Section 281 of the Patent Code: "A patentee shall have remedy by civil action for infringement of his patent." Suits for infringement follow the procedure of the federal courts. Suits are initiated in the United States district court in the district where the defendant resides or where he has a regular and established place of business and has committed the acts of infringement. From the decision of the district court, there is an appeal to the appropriate federal court of appeal.

One must be careful about making accusations of infringement unless he is ready to go to court for the following reason. If a patentee notifies another that he is infringing his patent or threatens suit, the patentee opens himself up to an affirmative action. By affirmative action, the one charged with infringement can initiate infringement proceedings in district court to get a judgment on the matter. Therefore, one should never accuse another with infringement unless he is ready to contest the matter.

The United States government may use any patented invention without permission of the patentee. The patentee is entitled to compensation from the government, however. When there is infringement by the government, damages are determined by the United States Court of Claims.

The presumption of validity of issued patents and defenses for infringement are set forth in Section 282 of the Patent Code cited below.

> A patent shall be presumed valid. Each claim of a patent (whether independent, dependent, or multiple dependent form) shall be presumed valid independently of the validity of other claims; dependent or multiple dependent claims shall be presumed valid even though dependent upon an invalid claim. The burden of establishing invalidity of a patent or any claim thereof shall rest on the party asserting such invalidity.
>
> The following shall be defenses in any action involving the validity of infringement of a patent and shall be pleaded:
>
> (1) Noninfringement, absence of liability for infringement, or unenforceability,
>
> (2) Invalidity of the patent or any claim in suit on any ground specified in part II of this title as a condition for patentability,
>
> (3) Invalidity of the patent or any claim in suit for failure to comply with any requirement of sections 112 or 251 of this title,
>
> (4) Any other fact or act made a defense by this title. In actions involving the validity or Infringement of a patent the party asserting invalidity or noninfringement shall give notice in the pleading or otherwise in writing to the adverse party at least thirty days before the trial, of the country, number, date, and name of the patentee of any patent, the title, date, and page numbers of any publication to be relied upon as anticipation of the patent in suit or, except in actions in the United States Court of Claims, as showing the state of the art, and the name and address of any person who may be relied upon as the prior inventor or as having prior knowledge of or as having previously used or offered for sale the invention of the patent in suit. In the absence of such notice, proof of the said matters may not be made at the trial except on such terms as the court requires.

During infringement litigation, two issues are generally involved. The patentee has the burden of establishing that his patent rights have been infringed, and the defense in return asserts that his activity does not constitute infringement, or he may assert that the patent he allegedly infringed is invalid. If the defense can prove invalidity, it follows that the patent was not infringed.

When the question of invalidity is raised, it is determined by the court. Since the patent is presumed valid when it is issued, the burden of establishing invalidity of the patent, or any claim thereof, rests on the party asserting invalidity. Each claim is presumed valid independently of the validity of other claims in the patent.

The defense may assert invalidity for any of a number of reasons. The most common reason is lack of invention for failure to meet the conditions of Section 102 and 103 of the Patent Code. As a general rule, when the prior art presented by the defense is the prior art of record that the examiner considered, the court usually rules in favor of the patentee. If the prior art presented is different and pertinent, the chances of invalidity are significantly greater. It is therefore important to have all pertinent prior art considered by the examiner during examination.

In addition, the defense may assert invalidity on informal grounds such as the patent being vague or naming the wrong inventor, or the defense may assert invalidity on statutory bars such as prior use or sale more than one year from the filing date as set forth in Section 102 of the Patent Code.

The decision of the court may be valid and infringed, valid but not infringed, invalid but not infringed, and invalid but infringed if valid. These have significance as far as appeals are concerned.

If the decision is valid and infringed, the patentee may obtain an injunction to prevent the continuation of the infringement, and he may ask for award of damages. The provision for an injunction is set forth in Section 283 cited below.

> The several courts having jurisdiction of cases under this title may grant injunctions in accordance with the principles of equity to prevent the violation of any right secured by patent, on such terms as the court deems reasonable.

Provisions for damages are set forth in Sections 284 and 285 cited below.

> Upon finding for the claimant the court shall award the claimant damages adequate to compensate for the infringement but in no event less than a reasonable royalty for the use made of the invention by the infringer, together with interest and costs as fixed by the court.
>
> When the damages are not found by a jury, the court shall assess them. In either event the court may increase the damages up to three times the amount found or assessed.
>
> The court may receive expert testimony as an aid to the determination of damages or of what royalty would be reasonable under the circumstances.

Damages are usually awarded after an accounting. This usually amounts to the profits which the infringer made during the infringement, but never less than a reasonable royalty. If the infringer willfully infringed the patent, he can be liable for attorney fees as set forth in Section 285: "The court in exceptional cases may award reasonable attorney fees to the prevailing party."

Section 286 sets forth a statutory time limit of six years prior to filing the complaint for infringement.

> Except as otherwise provided by law, no recovery shall be had for any infringement committed more than six years prior to the filing of the complaint or counterclaim for infringement in the action.

In the case of claims against the United States Government for use of a patented invention, the period before bringing suit, up to six years, between the date of receipt of a written claim for compensation by the department or agency of the Government having authority to settle such claim, and the date of mailing by the Government of a notice to the claimant that his claim has been denied shall not be counted as part of the period referred to in the preceding paragraph.

EMPLOYEE-EMPLOYER RIGHTS

Most of the research and development in the United States is done in an employee-employer relationship. This type of relationship exists in industrial concerns, academic institutions, the United States government, and consultant firms. All of the forenamed are employers and have certain rights as such.

Title to an invention must be in the name of the inventor. By statutory mandate, only the inventor may apply for a patent. Title 35, United States Code, Section 111, states in pertinent part: "Application for patent shall be made by the inventor, except as otherwise provided in this title." Title 35, United States Code, Section 116, states in pertinent part: "When an invention is made by two or more persons jointly, they shall apply for a patent jointly, and each sign the application." Should a patent issue naming the wrong inventor(s), the patent would be invalid.

According to these statutes, title to an invention must originate in the inventor. In the case of an employed inventor, however, title to the invention is subject to vestment to the employer by assignment unless there is an agreement to the contrary. This may result from an employment agreement, or it may result from common law in the absence of such agreement.

In most employee-employer relationships, there is a contractual agreement spelling out the rights of the employer to invention by assignment. Even in the absence of an expressed agreement, however, the law will imply an obligation to assign invention under most circumstances. While employment agreements are not in themselves an assignment, courts will require the employee to make assignment when the need arises.

The contractual agreement between employee-employer varies from company to company. These agreements usually contain an obligation to assign inventions, but some agreements contain special provisions. For example, some agreements limit the obligation for assignment to a particular technology. This type of agreement is more common in large companies with divisions involved in numerous areas of technology. By this special agreement, the employee is obligated to assign technology only in an expressed area. If he invents something in another technology, for example at home, he would not be obligated to assign by this agreement. Other agreements broadly define assignable inventions as any invention relating to the business of the company, while other agreements include all inventions. With the latter agreement, the employee is obligated to assign all inventions.

Some employment agreements contain additional elements. Such elements may require the employee to sign all documents relating to patent matters on his invention, or the agreement may contain elements requiring the employee to testify at any proceedings relating to his inventions. This gives the employer additional protection on his investment for invention when there is a conflict between employee-employer.

In the absence of an expressed agreement, the employer is still entitled to assignment of invention when the invention results from work the employee performs on the job. This vestment results from common law rights of an employer to an invention, and it has been consistently held by the courts. This law generally holds that when an employer hires a person for the purpose of making inventions, the employer is entitled to the fruit of the employee's creativity. This common law holds so long as there is no agreement otherwise. The employer can release the employee from this common law obligation, but this must occur in the form of a written agreement.

An application or a patent can be assigned. An assignment transfers the entire and undivided interest to the assignee. To constitute an assignment, the right to make, the right to use, and the right to sell must be transferred. If any of these rights are withheld, the agreement constitutes a license rather than assignment.

Section 261 of the Patent Code relates to ownership and assignment and is cited below.

> Subject to the provisions of this title, patents shall have the attributes of personal property.
>
> Applications for patent, patents, or any interest therein, shall be assignable in law by an instrument in writing. The applicant, patentee, or his assigns or legal representatives may in like manner grant and convey an exclusive right under his application for patent, or patents, to the whole or any specified part of the United States.
>
> A certificate of acknowledgement under the hand and official seal of a person authorized to administer oaths within the United States, or, in a foreign country, of a diplomatic or consular officer of the United States or an officer authorized to administer oaths whose authority is proved by a certificate of a diplomatic or consular officer of the United States, shall be prima facie evidence of the execution of an assignment, grant or conveyance of a patent or application for patent.
>
> An assignment, grant or conveyance shall be void as against any subsequent purchaser or mortgagee for a valuable consideration, without notice, unless it is recorded in the Patent and Trademark Office within three months from its date or prior to the date of such subsequent purchase or mortgage.

To legally pass title, the assignment must be in writing according to Section 261. The assignment itself is prima facie evidence of the execution of the assignment. The Patent Office will record assignments of patents and applications in accordance with Section 261, provided the submitted papers meet the requirements of Patent Office Rule 331 cited below.

(a) Assignments, including grants and conveyances, of patents, national applications, or international applications which designate the United States of America, will be recorded in the Patent and Trademark Office under 35 U.S.C. 261. Other instruments affecting title to a patent, a national application, or an international application which designates the United States of America, and licenses, even though the recording thereof may not serve as constructive notice under 35 U.S.C. 261, will be recorded as provided in this section or at the discretion of the Commissioner.

(b) No instrument will be recorded which is not in the English language and which does not amount to an assignment, grant, mortgage, lien, incumbrance, or license, or which does not affect the title of the patent or application to which it relates, except as ordered by the Commissioner.

(c) An instrument relating to a patent should identify the patent by number and date (the name of the inventor and title of the invention as stated in the patent should also be given); an instrument relating to a national application, or an international application which designates the United States of America should identify the application by serial number or international application number and date of filing (the name of the inventor and title of the invention as stated in the application should also be given), but if an assignment is executed concurrently with or subsequent to the execution of the application but before the application is filed or before its serial number or international application number and filing date are ascertained, it should adequately identify the application, as by its date of execution and name of the inventor and title of the invention; so that there can be no mistake as to the patentor application intended.

With an assignment, legal title passes whether or not the assignment is recorded with the Patent Office. There are, however, several important reasons for recording assignments. First of all, it protects the assignee. According to Section 261, if the assignment is not recorded in the Patent Office within three months from its date or prior to the date of a subsequent purchase or mortgage, the assignment is void against any subsequent purchaser or mortgagee for a valuable consideration without notice. The patentee can therefore, resell the patent if the assignment is not properly recorded.

The Patent Office allows public access to assignment records of issued patents. These records serve notice to subsequent purchasers that the patent has been assigned, thereby providing some protection to the assignee. Assignment records of patent applications are not open to public inspection, however, because of the confidentiality of applications.

Another reason for recording assignments is that it ensures the assignee the right to participate in the prosecution of the application, and it authorizes the Patent Office to issue the patent to the assignee in accordance with Section 152 of the Patent Code cited below.

Patents may be granted to the assignee of the inventor of record in the

Patent and Trademark Office, upon the application made and the specification sworn to by the inventor, except as otherwise provided in this title.

No special form is required for an assignment so long as it is in English and affects title. The Patent Office will provide an example of an assignment upon request.

APPENDIX:
SAMPLE PATENT

|||||| |||||||| ||| ||||| ||||| ||||| ||||| ||| |||| ||||| ||| ||||| ||||| |||||| |||

US005361216A

United States Patent [19]

Warn et al.

[11]   Patent Number: **5,361,216**

[45]   Date of Patent: **Nov. 1, 1994**

[54]   **FLOW SIGNAL MONITOR FOR A FUEL DISPENSING SYSTEM**

[75]   Inventors: **Walter E. Warn**, Knightdale; **Fred K. Carr**, Chapel Hill, both of N.C.

[73]   Assignee: **Progressive International Electronics**, Raleigh, N.C.

[21]   Appl. No.:   **907,548**

[22]   Filed: **Jul. 2, 1992**

[51]   Int. Cl.⁵ G06F 15/46; B67D 5/08
[52]   U.S. Cl. 364/510; 364/479; 364/178
[58]   Field of Search .... 364/509, 510, 551.01, 364/465, 178, 478, 479, DIG. 1, 229.41, 229.5; 73/3; 395/275, 325; 141/94, 98

[56]   **References Cited**

U.S. PATENT DOCUMENTS

4,250,550	2/1981	Fleischer	364/465
4,550,859	11/1985	Dow, Jr. et al.	222/26
5,132,923	7/1992	Crawford et al.	364/558
5,208,742	5/1993	Warn	364/131
5,270,943	12/1993	Warn	364/479
5,299,135	3/1994	Lieto et al.	364/479

FOREIGN PATENT DOCUMENTS

2600318   12/1987   France .

Primary Examiner—Thomas G. Black
Assistant Examiner—Michael Zanelli
Attorney, Agent, or Firm—Fred K. Carr

[57]   **ABSTRACT**

The present invention relates to a flow signal monitoring system for monitoring data signals in a data wire between fuel dispensers and the dispenser controller for collecting, storing, and later down-loading information relating flow quantity signals. The system includes an electronic communication translator which is attached to the wire. The design of the monitoring system is such that it is coupled to the data wire, however, it is electrically isolated from the data wire. When the dispenser and remote dispenser controller are communicating in current loop communication protocol, the system uses a configuration circuit with an opto-coupler having a light emitting diode and transistor for transforming the data signals into corresponding computer logic signals. When the dispenser and controller are communicating in voltage level communication protocol, a comparator is used to transform the data signals into corresponding computer logic signals. The computer logic signals are sent to a microprocessor with read-only memory and read-and-write memory for processing. The system further includes a data field selector which instructs the microprocessor to select and process data fields relating to flow quantity, and to discard all other data signals. Information on the amount of fuel dispensed at each fueling position in the dispenser is stored in memory. This information can be down-loaded to other devices including tank monitors, printer, and display devices.

18 Claims, 6 Drawing Sheets

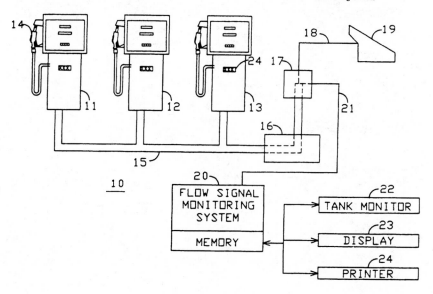

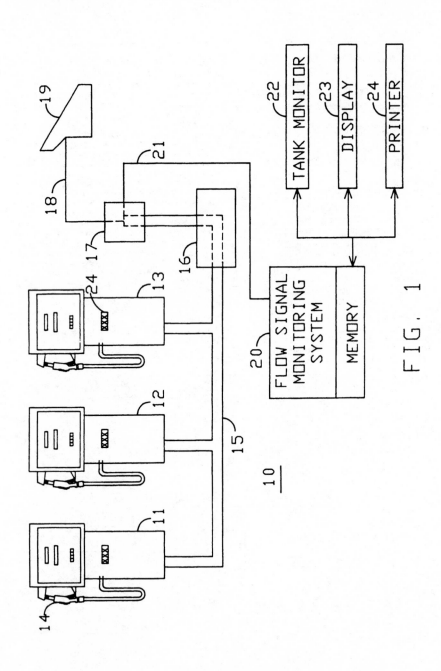

FIG. 1

U.S. Patent   Nov. 1, 1994   Sheet 2 of 6   5,361,216

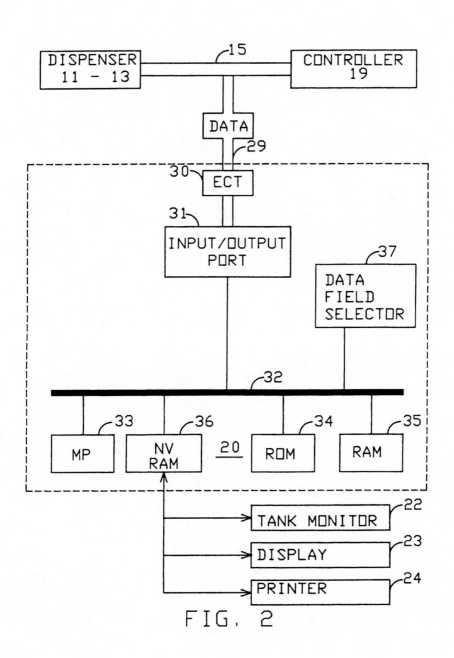

FIG. 2

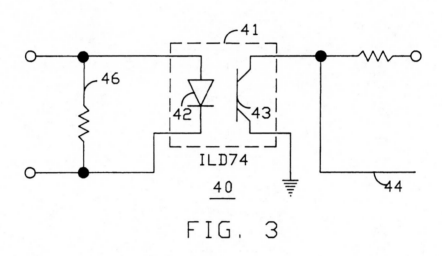

FIG. 3

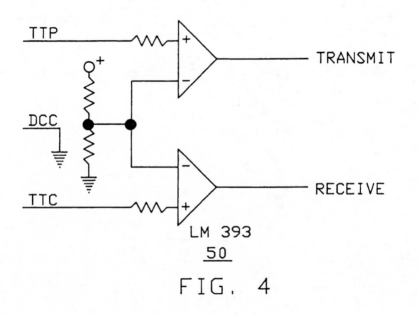

FIG. 4

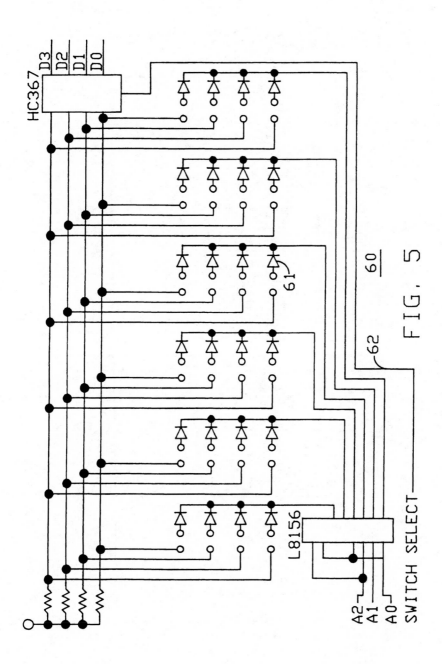

FIG. 5

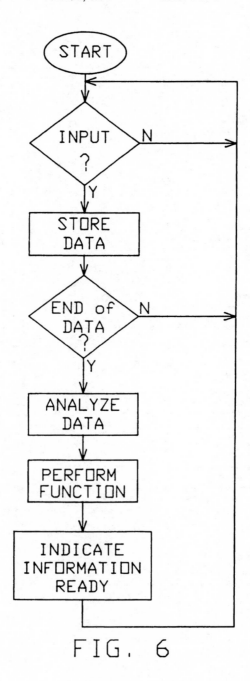

FIG. 6

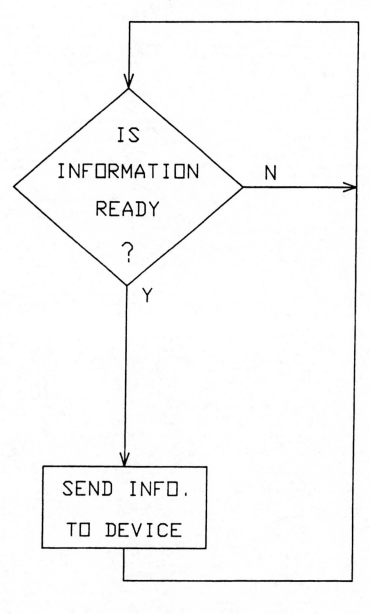

FIG. 7

APPENDIX: SAMPLE PATENT

5,361,216

1

FLOW SIGNAL MONITOR FOR A FUEL DISPENSING SYSTEM

FIELD OF THE INVENTION

The present invention relates to a device and method for monitoring data signals between fuel dispensers and a remote fuel dispenser controller, and in particular, data signals relating to the amount of fuel dispensed by the dispenser.

BACKGROUND OF THE INVENTION

To provide fuel for the traveling motorist, there are numerous fuel retail outlets located throughout. There has recently been a trend toward the motorist pumping his own fuel at so-called self service fueling sites which offer convenience and lower prices. The self service fueling sites most often have a fuel dispensing system where the dispensers are controlled by a remote dispenser controller. The remote dispenser controller may be located in a building at the site such that the dispensing process is controlled by a site attendant, or it may be a card read system whereby the customer controls the dispensing process.

Generally, a fuel dispenser includes a pump, a fuel supply pipe, a flowmeter, a flow quantity signal generator, a fuel supply hose with nozzle, and a flow indicator. The pump has at one end a pipe connection to a fuel supply tank, and at the other end a hose connection to a fuel supply nozzle. The flowmeter measures the quantity of fuel being pumped, and the flow quantity generator generates a flow quantity signal from the flowmeter. The indicator indicates the quantity of fuel being pumped based on the flow quantity signal.

As previously stated, the dispensers are often controlled by a remote dispenser controller located in a building at the fueling site, or through a card system. The remote controller has a wire connection between the dispensers and controller for transferring data signals. The remote controller generally is a microcomputer based system with read-only-memory (ROM), read-and-write memory (RAM), and input/output ports for reading and storing information applied at the ports. Specific functions of the control systems are well known in the art, and widely used in the industry. The microprocessor based control systems may be in the form of a stand alone console, or in the form of a logic module which interfaces the dispensers to a cash register system, or a card read system. The principles involved are the same, and it is understood that the present invention relates to all such systems.

The dispenser controller sends data signals to the dispensers, and the dispensers send data signals to the controller. Data signals sent to the dispenser from the controller include price per gallon to be charged at corresponding pumps, preset limits to be pumped, and pump authorization. Data signals sent from the dispenser to the controller include pump identity (pump number), pump status, and dispensed fuel volume and value.

In brief, the present invention relates to a flow signal monitoring system for monitoring flow quantity data signals in the data wire without interrupting data signal flow. The flow signal monitoring system is connected to the data wire between the dispensers and controller, and it monitors the data signals without interrupting the communication between the dispensers and controller. All data signals in the data stream are monitored, however, the data signals relating to flow quantity are selected and processed, and all other data signals are discarded. This is accomplished by a data field selector which designates data fields to be selected, and further instructs the microprocessor to select and process designated signals, and to discard all other signals. Information on dispensed volume is stored in memory and can be latter down-loaded to other devices including tank monitoring systems, printing devices, and visual display devices. These are, however, to be taken only as illustrative examples in that data extracted and stored can be used in other ways.

One use for the present invention is in combination with tank monitors. Recent Federal law requires that underground fuel storage tanks be continuously monitored to identify any loses caused by leaks. Tank monitors are used to do this, and are widely used in the industry for this purpose. Tank monitors use a probe which is permanently mounted in the storage tank through a riser pipe. Most tank monitor probes operate on a capacitance principle to sense fuel height. The probe has a wire connection to a microprocessor based control center which processes and stores the information, which usually includes gallons of fuel, inches of water, inches of fuel, temperature of fuel, and ullage.

Tank monitoring systems monitor the amount of fuel in the tank by a probe; the present invention collects and stores information on the actual amount of fuel dispensed from a tank. Thus, the present invention provides a method for reconciling the actual amount of fuel dispensed with the information collected by the tank monitor. The present invention stores actual transactions for each fueling position in memory, and this information can be retrieved at any time for comparison with information taken from the tank monitor.

Another use of the present invention is with inventory control and inventory report preparation. Each fueling position in a dispenser has a mechanical counter for counting the amount of fuel dispensed, and it keeps a running total of these amounts. These values are recorded and displayed on a numbered wheel in the dispenser. It is practice in the industry for the attendant to record these values at the end of his shift. These valves can then be used for shift totals, daily totals, inventory control, and related. For the site attendant to record these values, he must go out to the dispenser and visually observe the numbers from the display and write them down. This can be an inconvenience. For example, if the attendant is alone he must temporarily interrupt all dispensing while he is outside away from the controller. In cold climates, going outside can be uncomfortable. The present invention provides a device from which the site attendant can obtain these values from inside. These values can be visually observed, or they can be printed out.

SUMMARY OF THE INVENTION

In summary, the present invention relates to a flow signal monitoring system for monitoring data signals in a data wire between fuel dispensers and the dispenser controller without interrupting data signal flow in the wire. An electronic communication translator is attached to the wire to be monitored. The design of the monitoring system is such that it coupled to the data wire, however, it is electrically isolated from the data wire. When the dispenser and remote dispenser controller are communicating in current loop communication

5,361,216

<table>
<tr><td>3</td><td>4</td></tr>
</table>

protocol, the system uses a configuration circuit with an opto-coupler having a light emitting diode and transistor for transforming the data signals into corresponding computer logic signals. When the dispenser and controller are communicating in voltage level communication protocol, a comparator is used to transform the data signals into corresponding computer logic signals. The computer logic signals are sent to a microprocessor with ROM and RAM memory for processing. A data field selector switch instructs the microprocessor to select and process data fields relating to flow quantity, and to discard all other data signals. Information on the amount of fuel dispensed at each fueling position in the dispenser is stored in memory. This information can be down-loaded to other devices including tank monitors, printing devices, and display devices.

The primary object of the present invention is to provide a flow quantity signal monitoring system which monitors data signals in a data wire between fuel dispensers and remote dispenser controller without interrupting data signal flow in the data wire.

Another object of the present invention is to provide a flow signal monitoring system which can monitor data signals between fuel dispensers and remote dispenser controller for collecting and storing information on the amount of fuel dispensed at each fueling position.

A further object of the present invention is to provide a flow signal monitoring system which can be used in combination with a tank monitoring system for reconciling the actual fuel volume dispensed with the information collected by the tank monitor.

A further object of the present invention is to provide a flow signal monitoring system which can be used in combination with a printer for printing the volume of fuel dispensed at each fueling position.

A further object of the present invention is to provide a flow signal monitoring system which can be used in combination with a display device for displaying the volume of fuel dispensed at each fueling position.

Other objects of this invention will appear in the following specification and claims, reference being made to the accompanying drawings which form a part thereof.

BRIEF DESCRIPTION OF THE DRAWINGS

FIG. 1 is a schematic view of a dispensing operation including dispensers, dispenser controller with attachment of the flow signal monitoring system incorporating the principles of the present invention.

FIG. 2 is a schematic depiction of the components of the flow signal monitoring system with connection to other devices to which stored information can be downloaded.

FIG. 3 is a schematic diagram of an opto-coupler as used in the configuration circuit of the present invention.

FIG. 4 is a schematic diagram of a comparator as used in the configuration circuit of the present invention.

FIG. 5 is a schematic diagram of the data field selector for designating data fields in the present invention.

FIG. 6 is a flow chart demonstrating the overall system processing for the flow signal monitoring system.

FIG. 7 is a flow chart for demonstrating the downloading of information from memory to another device.

DETAILED DESCRIPTION OF THE INVENTION

Referring now to the drawings, and first to FIG. 1, there is shown a schematic view of a fuel dispensing operation, generally designated (10), with three electronic fuel dispensers (11-13) connected to a remote dispenser controller (19). The dispensers (11-13) and dispenser controller (19) are electrically connected by a data wire (15). The dispenser controller (19) controls the fuel dispensing process at the dispensers (11-13). The data wire (15) is often enclosed in an underground wiring trough (16) for protection. In the illustration, an example of three dispensers is used, however, it is common in the industry to have from two to sixteen dispensers with multiple fueling positions at a fueling site which are controlled by a remote controller. The three dispensers are used for illustration only. It is understood that the present invention has application in any number of multi-product dispensers.

Several types of dispenser control systems are presently used to control electronic and electro-mechanical dispensers, including stand alone consoles, electronic cash registers interfaced to the dispensers through a logic module, and card read systems. Typically on these systems there are a series of input buttons (not shown) on the dispenser controller for entering information on the dispensing process, and a display for displaying information during the dispensing process. Generally, information transferred between the dispensers and controller is in the form of digital data signals. Data signals from the controller (19) to the dispensers (11-13) include price per gallon to be charged at the dispensers, preset amounts of fuel to be dispensed, and pump authorization (i.e., an activated mode where the pump will dispense fuel when the customer opens a valve in the pump nozzle). Data signals sent from the dispensers (11-13) to the controller (19) include pump number, pump status, and dispensed fuel volume and value for each fueling position.

Referring further to FIG. 1, it can be seen that the flow signal monitoring system FSMS (20) is attached to the data wire (15) such as to monitor the data signals. In the illustration, the dispensers (11-13) and the dispenser controller (19) are communicating in current loop, and the FSMS (20) is connected to the data wire (15) through connection (21) in the distribution box (17). When the dispensers and controller are communicating in voltage level, later discussed, the connection is in the controller cable (18). The FSMS (20) monitors the communication, i.e. data stream, between the controller (19) and dispensers (11-13) without interfering with the flow of data signals in the wire (15). In essence, the FSMS (20) is coupled to, but electrically isolated from, the data wire (15) through a configuration circuit, later discussed. Data signals relating to dispensed fuel volume are selected, processed, and stored in memory, other data fields are discarded.

During a dispensing operation, a customer pulls his vehicle along side one of the dispensers and removes the nozzle (14) from the dispenser. The dispenser is authorized, or enabled, by the dispenser controller; enabling occurs by activating relay contacts in the pump. When the pump is activated, it dispenses fuel when the customer opens the valve in nozzle (14). Each fueling position in a dispenser has a mechanical counter for counting the amount of fuel dispensed, and it keeps a running total of these amounts. The values are recorded by a

5,361,216

5

mechanical totalizer and displayed on a numbered wheel (24) in the dispenser.

As fuel is dispensed, a pulse signal is generated by a pulser in a conventional manner. The pulses are transferred to a processing unit (not shown) in the dispenser head, which is connected to the dispenser controller (19) through data wire (15). The FSMS (20) monitors the fuel flow quantity signal as it is being transmitted from the dispenser to the controller. This information is stored in memory, and can be later down-loaded to other systems including a tank monitor (22), a display device (23), and printing device (24).

Referring now to FIG. 2, there is shown a diagram of the components of the FSMS, generally designated (20), with attachment to the data wire (15). In general, the FSMS (20) includes an electronic communication translator ECT (30), a microprocessor MP (33) with read-only-memory ROM (34) and read-and-write memory RAM (35), a data field selector (37), and a memory device (36) for storing information on the quantity of fuel dispensed. These are commonly connected through common data bus (32).

The ECT (30), which is in essence, a configuration circuit, is connected to the data wire (15) between the dispensers (11-13) and dispenser controller (19); the connection may be in series or parallel. The ECT (30) allows the data current to monitored without interrupting communication between the dispensers and controller. The ECT (30) couples the data wire to the FSMS (20), while at the same time it electrically isolates the two. The ECT (30) translates the dispenser-controller digital data signals into computer logic signals which are sent to the MP (30) for reading. The ECT (30) is connected to the MP (33) through an input/output port (31). The input-output port includes URAT chips for sending interrupt signals through bus connection (32) to MP (33).

The MP (33) in the FSMS (20) operates in a conventional manner. Specific implementations of the MP (33) are well known to those skilled in the art, and include for example, integrated circuits manufactured and sold by INTEL (model 9135 KC). The MP (33) is functionally connected to a ROM chip (34), and RAM chip (35). Program control for the MP (33) is stored in ROM, and is set forth in flow chart form in FIG. 6. The computation programs are stored in RAM, and is set forth in flow chart form in FIG. 6. Once the information on the amount of fuel dispensed at each fueling position has been processed by the MP (33), it can be stored generally in any storage device. In the illustrative example, storage is in a non volatile read-and-write memory chip NVRAM (36) such that stored information will not be lost if power is lost. The information can also be stored in the system operating RAM (35), or it could be transmitted to storage devices exterior to the FSMS (33) circuit board.

A data field selector (37) is connected to the MP (33) through bus (32). As previously discussed, the data stream presented to the MP (33) include signals from the dispenser controller to the dispensers including price per gallon, preset limits of fuel to be dispensed, and pump authorization; data signals from the dispensers to the controller include pump number, pump status, and the fuel dispensed. The data stream may be in byte protocol form or bit protocol form. The data field selector (37), in essence, provides a method for parsing the data stream. The data field selector (37) instructs the MP (33) as to which data fields to process, and which

6

data fields to discard. In the present invention, the data field selector board (60), seen in FIG. 5, instructs the MP (33) to select and process computer logic fields which correspond to data fields relating to fuel flow, and to discard all other data fields.

Different commercial dispenser manufacturers use a unique communication protocol for communication between their dispensers and controller. The two most widely used are current loop and voltage level. With current loop, there is a wire running from the distribution box to each dispenser and back to the distribution box forming a loop (serial topographic arrangement). An example is shown in FIG. 1. With voltage level, there is a wire pair running from a site controller to each dispenser, i.e., a parallel topographic arrangement.

An advantage of the present invention is that it can monitor the data signals in different commercial brands of dispensers, and different dispenser models within a commercial brand. This is done by a configuration circuit in the ECT (30). For this discussion, illustrative examples of current loop and voltage level communication are used. It is understood that the configuration circuit in the ECT (30) can be adapted for other types of communication between the dispensers and controller.

Referring now to FIG. 3, there is shown a schematic diagram of a current loop configuration circuit, generally designated (40), as used in the ECT (30) when the dispensers and controller are communicating in current loop. The configuration circuit (40) converts the dispenser-controller digital data signals into computer logic signals which are presented to the input-output port (31), and then to the MP (33). The configuration circuit (40) includes an opto-coupler (41) which is connected to the data wire (15) for monitoring the data signals. The configuration circuit (40) has a resistor (46) which allows one to adjust the amount of current flowing into the circuit. The opto-coupler converts the data signals into a computer logic data stream which is readable by the MP (33), for example, transistor-transistor logic TTL signals. The computer logic data stream corresponds to the digital data stream in the data wire. The opto-coupler (41) samples the data flow through the light emitting diode (42), and this information is transferred to the transistor (43). The transistor (43) generates signals in, for example, TTL form for presentation to the MP (33) through input-output port (31), shown in FIG. 2. During operation, the transistor (43) may, for example, apply five volts to the MP (33) through connection (44) for a high signal bit, and zero volts for a low bit signal; other voltages can be used. The configuration circuit (40) has a conventional band rate output chip, not shown, with connection to the input-output port (31) for synchronizing signal flow. Depending on the hardware arrangement, other types of computer logic including CMOS, NMOS, and related can be used.

The design of the communications translator depends on the type communication between the dispensers and controller. Referring now to FIG. 4, there is shown a comparator, generally designed (50), as used in the ECT (30) when the dispensers and dispenser controller use voltage level communication. With voltage level, the dispenser-controller communicate through voltage differential in a wire pair running from a site controller to each dispenser. Commercially available model LM 393 is an illustrative example of a comparator as used in the present invention.

5,361,216

7

Referring now to FIG. 5, there is shown the data field selector board, generally designated (60). As previously discussed, there are several data signal fields flowing between the dispensers (11-13) and dispenser controller (19). These signals are converted by the ECT (30) into computer logic signals for presentation to the MP (33). The data field selector (60) understands the communication protocol of the data signals, and is programmed to cause the MP (33) to parse the computer logic data stream such as to select and process certain data field, and to discard all other data fields. The data field selector (60) is a matrix of diodes, as example (61), forming in essence a configuration board. The data field selector board (60) communicates with MP (33) through an address bus (A-A2), and a reader bus (D-D3). In the present invention, the dispenser-controller data signal fields selected and processed are pump number, pump status, and the amount of fuel dispensed by these, which in combination form the flow quantity signal. The data field selector (60) causes the MP to parse the data stream selecting these fields, and discarding all other fields. The MP (33) is further connected to the data field selector board (60) through switch select (62).

Referring now to FIGS. 6 and 7, there are shown flow charts for program processing contained in ROM (34) and RAM (35) of the FSMS (20). When data is presented, the data is stored until the end of data presentation. Thereafter, the data is analyzed and the function performed. As an example, a customer pulls up to a dispenser, for illustration, say fueling hose designated number 3. He removes the nozzle, and inserts it in the fuel tank. The dispenser is authorized, and the customer squeezes the nozzle trigger. When the fuel starts to flow, the dispenser sends to the dispenser controller the following; hose number 3 (pump number) is pumping (pump status) along with a continuous update on the amount dispensed, which forms the flow quantity signal. When the customer has finished pumping, a signal is send indicating that hose number 3 is now idle. The total amount dispensed is calculated, and put in memory (totalizer) such that there is a running total of fuel dispensed by hose number 3. The same information is kept for each hose (fueling position) in the fueling network. At request this information can be taken from memory and transferred to the device requesting the information, or the memory can be associated with a clock such that the information is down-loaded at pre-programmed times.

During operation, the stored information on fuel dispensed is best down-loaded when all dispensers are idle. This provides a "snap shot" of all activity at each fueling position on a real time basis. Referring to FIG. 7, there is shown a flow chart for this. When the information is requested, or is to be down-loaded at a preset time, the information is transmitted under these conditions.

Having discussed above the operation of the flow signal monitoring device (20), attention is now directed toward examples of usage at a fueling facility. An important feature of the present invention is that it can be used with different commercial brands of dispensers communicating in different communication protocols. By monitoring the data signals, actual fuel transactions are extracted and totals for each fueling position stored in memory. This information can be displayed by a display device mounted on the FSMS (20) housing, or it can printed through a printer port coupled to the system. In addition, it can be transmitted through a com-

8

munication interface, as examples RS-232 and RS-485 ports, to an on-site computer at the fueling facility, or over telecommunications lines to a host computer at headquarters. In essence, once this information has been collected and stored, there are a number of existing technologies which would allow the information to be used on site, or transmitted to remote locations.

As previously discussed, tank monitors are widely used in the industry to monitor fuel storage tanks by a probe. The present invention can be used in combination with a tank monitor to reconcile the actual amount of fuel dispensed from a tank with the information collected by the tank monitor. Presently existing tank monitors have either: a processing unit built-in, are associated with a separate computer at the fueling site, or are coupled to host computing system at a remote location through a telecommunications lines. It is understood that the present invention can down-load information to each of these arrangements. RS-232 and RS-485 ports can be used for interfacing the FSMS (20) to the tank monitor. In addition, the information can be down-loaded at request, or the memory device can be associated with a clock (commercially available from Dallas Semiconductor) for causing the information to be transmitted at a pre-set time.

As also previously discussed, each fueling position in a fuel dispenser has a mechanical counter for keeping a running total of fuel dispensed. There is a mechanical display, or numbered wheel, in each dispenser for displaying the amount of fuel dispensed from each fueling position. The totals are from when the display was first set to zero, i.e., a running total. These totals are widely used in inventory control. It is common practice to have the site attendant write down these totals at the end of his shift for inventory and record keeping. To do this, he must go outside to the dispensers, and write down the numbers.

The present invention can be used to collect and store the same information from inside. This information can be displayed or printed from inside the facility, or it can be transmitted over communication lines to remote host computers. When the FSMS (20) is installed in an existing facility, the memory in the FSMS (20) can be set to correspond with the existing values in the mechanical totalizer in the dispenser. Thereafter, the numbers displayed or printed from the FSMS (20) are the same as those in the dispenser, thereby, providing a method for obtaining these numbers without going outside. The values in the memory of the FSMS (20) can be set by an input keypad coupled to the MP (33). As stated above, these numbers are set to correspond to those in the dispenser when the FSMS (20) is installed.

In one embodiment of the present invention, a liquid crystal display and input key pad are mounted on the top of box housing the flow signal monitoring system circuit boards. Liquid crystal displays and keypad are widely used for other applications, and commercially available. The commercially available liquid crystal is coupled to the MP (33) in the FSMS (20), and displays totals from memory when activated. The system can also be equipped with a printer port for coupling to a printer. The information in memory is down-loaded through the port to the printer when a hard copy is desired.

The above described invention relates to a method and device for monitoring data signals in a data wire between a dispenser and dispenser controller such that information on the amount of fuel dispensed is extracted

5,361,216

9

and stored. While the invention has been described in the manner presently conceived to be most practical and a preferred embodiment thereof, it will be apparent to persons ordinarily skilled in the art that modifications may be made thereof within the scope of the invention, which scope is to be accorded the broadest interpretation of the claims such as to encompass all equivalents, devices, and methods.

What is claimed is:

1. A flow signal monitoring system having a microprocessor for monitoring a digital data stream in a data wire between at least one fuel dispenser and a dispenser controller, wherein said data stream includes a plurality of data fields including one data field comprising a command code, pump number, pump status, and dispensed fuel volume and a value containing information on the amount of fuel dispensed from each fueling position which is collected and stored for down-loading to a display, a printer, or tank monitor; comprising:

    (a) an electronics communication translator means, connected to said data wire between said at least one fuel dispenser and said dispenser controller, for converting said data stream into a computer logic data stream having corresponding data fields including information on the amount of fuel dispensed from each fueling position;

    (b) a microprocessor including plural input-output ports, programmable read only memory, programmable read and write variable memory, said microprocessor having bus means connecting said communication translator with at least one of said microprocessor ports for receiving said computer logic data stream;

    (c) a programmable data field selector means, having bus connection to at least one of said microprocessor ports, for instructing said microprocessor to recognize, select and process said corresponding data fields containing information on the amount of fuel dispensed from each fueling position, and to discard all other data fields; and

    (d) a memory means, coupled to said microprocessor, for receiving and storing information on the amount of fuel dispensed from each fueling position.

2. A flow signal monitoring system as recited in claim 1, wherein: said communication translator means uses an opto-coupler means including a light emitting diode and transistor for converting said data stream into said computer logic data stream.

3. A flow signal monitoring system as recited in claim 1, wherein: said communications translator means uses a comparator means for converting said data stream into said computer logic data stream.

4. A flow signal monitoring system as recited in claim 1, further comprising: a communication interface means, coupled to said microprocessor and said memory means, for down-loading stored information on the amount of fuel dispensed from each fueling position to said tank monitor.

5. A flow signal monitoring system as recited in claim 1, further comprising: a communications interface means, coupled to said microprocessor and said memory means, for down-loading stored information on the amount of fuel dispensed from each fueling position to said printer.

6. A flow signal monitoring system as recited in claim 1, further comprising: a communications interface means, coupled to said microprocessor and said mem-

10

ory means, for down-loading stored information on the amount of fuel dispensed from each fueling position to said display.

7. A flow signal monitoring system as recited in claim 6, wherein: said display is a liquid crystal display means for displaying the amount of fuel dispensed.

8. A microprocessor controlled method for collecting and storing information on an amount of fuel dispensed from each fueling position in at least one fuel dispenser, utilizing a data line monitor to monitor a data stream in a data wire between said at least one fueld dispenser and a dispenser controller wherein said data stream includes a data field comprising command code, pump number, pump status, and dispensed fuel volume and a value containing information on the amount of fuel dispensed form each fueling position, comprising the steps of:

    (a) feeding the data stream in the data wire between said at least one fuel dispenser and dispenser controller into an electronics communication translator means for converting said data stream into a corresponding computer logic data stream including a computer logic data field containing information on the amount of fuel dispensed from each fueling position;

    (b) feeding said computer logic data stream to an input-output pin of a microprocessor having programmable read only memory and program/nable read and write memory;

    (c) coupling said microprocessor to a data field selector means for instructing said microprocessor to recognize, select and process said computer logic data field containing information on the amount of fuel dispensed from each fueling position, and to discard all other data fields;

    (d) storing information on the amount of fuel dispensed from each fueling position in a memory means for keeping track of the amount of fuel dispensed from each fueling position in said at least one fuel dispenser;

    (e) down-loading said information on the amount of fuel dispensed from each fueling position from said memory means through a communications interface means to a second device for transferring said information so that said second device now has information on the amount of fuel dispensed from each fueling position in said at least one fuel dispenser.

9. The method as recited in claim 8, wherein: step (a) is practiced by using an opto-coupler means in said communications translator means for converting said data stream into said computer logic data stream.

10. The method as recited in claim 8, wherein: step (a) is practiced by using a comparator means in said communications translator means for converting said data stream into said computer logic data stream.

11. The method as recited in claim 8, wherein: step (e) is practiced by down-loading said information on the amount of fuel dispensed form each fueling position through said communications interface means to a computing element in said tank monitor for reconciling the actual amount of fuel dispensed from each fueling position with information collected by said tank monitor.

12. The method as recited in claim 8, further comprising the step of: coupling a programmable clock means to said microprocessor and said memory means for instructing step (e) to be carried out at a pre-programmed time.

5,361,216

11

13. The method as recited in claim 8, wherein: step (e) is practiced by down-loading said information on the amount of fuel dispensed from each fueling position through said communications interface means to a liquid crystal display means for displaying said information.

14. The method as recited in claim 8, wherein: step (e) is practiced by down-loading said information on the amount of fuel dispensed from each fueling position through said communications interface means to a printer for printing said information.

15. A method for operating a microprocessor for collecting, storing, and transmitting information on an amount of fuel dispensed from a fuel dispenser with a flow signal monitoring system, connected to a data wire between the fuel dispenser and a dispenser controller wherein said data wire conducts a data stream including data fields with information on the amount of fuel dispensed from each fueling position on said fuel dispenser, including a communications translator for converting the data fields in said data stream in said data wire into a computer logic data stream including computer logic data fields with information on the amount of fuel dispensed from each fueling position and a data field selector for instructing the microprocessor to select and process said computer logic data fields with information on the amount of fuel dispensed from each fueling position and to discard all other data fields, wherein the information on the amount of fuel dispensed is downloaded to a tank monitor, comprising the steps of:

    (a) receiving from said communication translator said computer logic data stream corresponding to the data stream in said data wire between said dispenser and said dispenser controller;

    (b) receiving from said data field selector a selector signal including instructions to select data fields with information on the amount of fuel dispensed from each fueling position;

    (c) parsing said computer logic data stream for selecting and processing data fields with information on the amount of fuel dispensed from each fueling position, and discarding all other data fields;

    (d) storing the information on the amount of fuel dispensed from each fueling position in a memory chip.

16. A method of operating a microprocessor for collecting, storing, and transmitting information on the amount of fuel dispensed as recited in claim 15, further comprising the step of: transmitting the stored information on the amount of fuel dispensed from each fueling position from said memory chip through a communications interface to said tank monitor.

17. A method of operating a microprocessor for collecting, storing, and transmitting information on the

12

amount of fuel dispensed, as recited in claim 15, further comprising the step of: transmitting the stored information on the amount of fuel dispensed from each fueling position at pre-programmed times, wherein said memory chip is coupled to a programmable clock means for instructing the stored information to be down-loaded through a communications interface at pre-programmed times, to said tank monitor.

18. A flow signal monitoring system having a microprocessor, used in combination with a tank monitor, for monitoring data signals in a data wire between at least one fuel dispenser and a dispenser controller without interrupting data signal flow in the data wire, where said data wire conducts a data stream including information on the amount of fuel dispensed form each fueling position in said at least one fuel dispenser which is collected, stored, and later down-loaded to the tank monitor for reconciling the amount of fuel dispensed with the information collected by the tank monitor, comprising:

    (a) an electronics communications translator means, connected to said data wire, for translating the data signals in said data stream into a computer logic data stream having data fields corresponding to said data signals in said data wire such that data signal flow in said data wire is not interrupted;

    (b) a microprocessor having plural input-output ports, programmable read and write only memory, programmable read and write variable memory, said microprocessor having a bus connecting said communication translator with at least one of said microprocessor ports for receiving said computer logic data stream;

    (c) a programmable data field selector means having a bus connection to at least one of said microprocessor ports for instructing said microprocessor as to which of said data fields in said data stream to select and process, and which of said data fields to discard, wherein said data field selector means instructs said microprocessor to parse said computer logic data stream for selecting and processing data fields with information on the amount of fuel dispensed from each fueling position, and to discard all other data fields;

    (d) a memory means having a bus connection to at least one of said microprocessor ports for storing information on the amount of fuel dispensed from each fueling position;

    (e) a communications interface means, coupled to said memory means and further coupled to said tank monitor, for down-loading said information on the amount of fuel dispensed from each fueling position at request to said tank monitor.

• • • • •

INDEX